Transformative Language Teaching for Sustainability

"This publication offers a transformative approach to embedding sustainability in language teaching, providing valuable insights both for researchers with an in-depth knowledge of the field and practitioners who address sustainability in their classrooms, curricula, and institutions and wish to gain a greater insight into the underlying research and theory.

A diverse group of experts have written chapters that give the book a global span and critical edge. The opening theoretical section courageously problematises core concepts that we may take for granted—sustainability, education for sustainability and the sustainable development goals. It challenges the reader to re-examine their current beliefs and practices and argues the case for the centrality of language teaching in education for sustainability. Practical application chapters range from German students of Swedish problematising the green energy transition as they investigate the impacts of such projects on the indigenous Sámi people of Northern Sweden, to addressing linguistic and environmental justice in Colombia, offering inspiring case studies for language teaching professionals."

—Owain Llewellyn, *Bangor University International College, UK and Founder,* ELTsustainable.org

Minna Maijala · Salla-Riikka Kuusalu ·
Riikka Ullakonoja
Editors

Transformative Language Teaching for Sustainability

From Theory to Practice

palgrave
macmillan

Editors
Minna Maijala
School of Languages and Translation Studies
University of Turku
Turku, Finland

Riikka Ullakonoja
School of Languages and Translation Studies
University of Turku
Turku, Finland

Department of Language and Communication Studies
University of Jyväskylä
Jyväskylä, Finland

Salla-Riikka Kuusalu
School of Languages and Translation Studies
University of Turku
Turku, Finland

Department of Biology
University of Turku
Turku, Finland

ISBN 978-3-031-85492-7 ISBN 978-3-031-85493-4 (eBook)
https://doi.org/10.1007/978-3-031-85493-4

Cover credit: Salla-Riikka Kuusalu

This Palgrave Macmillan imprint is published by the registered company Springer Nature Switzerland AG
The registered company address is: Gewerbestrasse 11, 6330 Cham, Switzerland

Acknowledgements

We are extremely grateful for the participation of students, pre-service teachers, teachers, scholars, schools, and universities in our studies.

This work was supported by the Kone Foundation (project Ethical and Sustainable Language Teaching, EKKO, grant number 202006277). The following authors of this volume have been members of the EKKO project: Leena Maria Heikkola, Salla-Riikka Kuusalu, Päivi Laine, Minna Maijala, Soila Merijärvi, Maarit Mutta, Mareen Patzelt, and Riikka Ullakonoja.

Contents

Notes on Contributors

Dr. Marianne Broermann teaches Swedish at the Department of Scandinavian Studies and the Centre for Languages and Transferable Skills at the University of Göttingen, Germany. She is particularly interested in sustainability in higher education and exploring which course topics and teaching methods can contribute to the sustainability of language courses and, thus, enhance students' learning of future skills.

Friederike Delius, MA studied linguistics at the University of Marburg, Germany. In the fall of 2023, she became a research assistant at Åbo Akademi University in Turku, Finland. Among her research interests are structural aspects of language and their implications for society.

Dr. Niklas Gericke is a professor of science education and the director of the research centre SMEER (Science, Mathematics, Engineering Education Research) at Karlstad University, Sweden, and a visiting professor at Inland Norway University of Applied Sciences, Norway. His main research interests are biology education and sustainability education from conceptual, teaching as well as implementation perspectives. Currently he is engaged in the pan-European project Impact Assessment for Action Competence (IMP>ACT).

Dr. Leena Maria Heikkola is a professor of Finnish and Kven language at UiT—Arctic University of Norway in Norway, and an adjunct professor of Finnish language at Åbo Akademi University in Finland. Her research focuses on Finnish didactics, inclusion and sustainability in language teaching, and clinical linguistics.

Dr. Attila Krizsán is an associate professor and university lecturer of English language at the University of Turku, Finland. His research focuses on language and identity, language and politics, and language and the environment.

Dr. Salla-Riikka Kuusalu has a background in ecological research, with a recent focus on ecolinguistic research and interdisciplinary research on sustainability education in language teaching. She is currently a postdoctoral researcher at the School of Languages and Translation Studies and a university teacher at the Department of Biology at the University of Turku, Finland.

Dr. Päivi Laine is a university lecturer of Finnish at the School of Languages and Translation Studies at the University of Turku, Finland. Her research focuses on learning and teaching Finnish as a first language. She is especially interested in the ability of pre-service language teachers to deal with multilingual and multicultural learners. Dr. Laine is also interested in students' meta-linguistic skills and aspires to develop language teacher education.

Dr. Minna Maijala (Ph.D., Ed.D.) is a professor of language teaching and learning at the School of Languages and Translation Studies at the University of Turku, Finland. Her areas of expertise include research on the teaching and learning of foreign languages, culture teaching, language teaching materials, language teacher education, and sustainability in language teaching and learning. She is the principal investigator of two projects: Ethical and Sustainable Language Teaching (KONE Foundation, 2021–2025) and Implementation of Sustainable Development Goals in Language Teaching and Learning (Research Council of Finland, 2024–2028).

Dr. Minna Maunula is a university lecturer and operations director at the University of Jyväskylä and Kokkola University Consortium Chydenius in Finland. Her areas of interest include the life course of adults, the relevance and sustainable future of education in a global era, and the development of web-based and multi-modal education in adult academic education.

Dr. Minna Maunumäki is a university teacher in education, adult education, and early childhood education at the University of Jyväskylä Open University and Kokkola University Consortium Chydenius in Finland. Her areas of interest include learning and learning assessment, education policy, and the development of web-based and multi-modal education in adult education.

Soila Merijärvi, MA is a project researcher at the School of Languages and Translation Studies at the University of Turku, Finland. She is a member of the research group EKKO (Ethical and Sustainable Development in Language Teaching).

Dr. Santiago Sánchez Moreano is a Marie-Curie research fellow at the French Institute of Research for Development (Centre for Evolutionary and Functional Ecology, CEFE), France. He is interested in the construction of social identities as well as the relationality between language, social actors, and nature in Amazonia in the construction of commonalities. He is the lead of the project Multilingualism, Sustainability, and Education in the Amazon Trapezium (Multi-SEAT), funded by Horizon Europe.

Mareen Patzelt, MA is a teacher of German and Finnish as a foreign language. She is currently working as a university teacher of German at the School of Languages and Translation Studies at the University of

Turku in Finland. As a Ph.D. student in German, she explores reading strategies in foreign language learning. She is also interested in language awareness and global citizenship education.

Dr. Meameno Aileen Shiweda is a university lecturer at the University of Namibia, Namibia, and specialist in English language education. Her research interests include teacher education, teacher identity, multilingualism, second language learning, intercultural communication, code-switching, and sustainability in both language education and higher education.

Dr. Arran Stibbe is a professor of ecolinguistics at the University of Gloucestershire, UK, and the founder of the International Ecolinguistics Association. He has also been awarded the prestigious National Teaching Fellowship for his significant contributions to student success in higher education. His research focuses on how language encodes and defines our relationship with ourselves, other animals, and our shared planet.

Dr. Riikka Ullakonoja is a researcher at the School of Languages and Translation Studies at the University of Turku and a member of the EKKO (Ethical and Sustainable Language Teaching) research group. She also works as a researcher at the University of Jyväskylä, Finland, in the AASIS (Automatic Assessment of Spoken Interaction in Second Language) project.

List of Figures

List of Tables

1

Introduction—Towards Transformative Language Teaching for Sustainability

Minna Maijala, Salla-Riikka Kuusalu,
and Riikka Ullakonoja

Introduction

Sustainable development is a dynamic process of striving for a sustainable future through different means (Sinakou et al. 2019). The Brundtland Report first introduced the concept of sustainable development to a wider public and defined it as 'development that meets the needs of the present without compromising the ability of future generations to meet their own needs' (WCED 1987, p. 43). The concept of sustainable

M. Maijala (✉) · S.-R. Kuusalu · R. Ullakonoja
School of Languages and Translation Studies, University of Turku, Turku, Finland
e-mail: minmai@utu.fi

S.-R. Kuusalu
Department of Biology, University of Turku, Turku, Finland

R. Ullakonoja
Department of Language and Communication Studies, University of Jyväskylä, Jyväskylä, Finland

M. Maijala et al. (eds.), *Transformative Language Teaching for Sustainability*,
https://doi.org/10.1007/978-3-031-85493-4_1

development is understood differently in different contexts (Wals 2009), but its focus is usually on biodiversity or global health (Murillo-Vargas et al. 2020). Although often associated with the physical environment, sustainable development encompasses broader themes of equality and social justice, which are at the core of language teaching (Zygmunt 2016; de la Fuente 2021). Research literature in general makes a distinction between the concepts of 'sustainable development' and 'sustainability'. The term 'sustainability' is typically associated with the ecological or environmental dimension of sustainable development, or the process of achieving the Sustainable Development Goals (SDGs). Consequently, sustainable development can also be understood as the ultimate objective of this process (Holden et al. 2014). As these two concepts are frequently used synonymously and pursue the same objectives, they are used alternatively in this chapter.

The SDGs are interlinked and have a common underlying influence in all subjects taught in schools within the UN member states (WCED 1987). As the themes of the 17 SDGs identified by UNESCO (2018, 2021) are part of all learning, they should also be included in language teaching and learning. The main tool to implement the SDGs and to advance the goals of UNESCO's (2018) Agenda 2030 action plan is its concept of Education for Sustainable Development (ESD). ESD is a context-specific, multidisciplinary, and learner-centred approach that encourages students to adopt a holistic way of thinking and enables change agency (Lewis et al. 2014; Annan-Diab and Molinari 2017; Sinakou et al. 2019; see also Chapter 2 in this volume). Nevertheless, the practical implementation of ESD in classrooms is often teacher-centred (Eilam and Trop 2010). Furthermore, teachers may perceive that existing school curricula and organisational structures impose certain constraints on the implementation of ESD (Fazio and Karrow 2013; Ntona et al. 2024). Recently, research on ESD competencies has flourished in many fields (Frisk and Larson 2011; Lozano et al. 2017), including studies on pedagogical approaches to foster sustainability education in language classrooms in the US (de la Fuente 2021; Goulah and Katunich 2020; Melin 2019). However, the impact of these approaches on learners' sustainability competencies (Lozano et al. 2017) and the efficacy of the methods used to develop these competencies have not been widely examined. In the field of language education, research on the implementation and effects of sustainable development remains limited (Micalay-Hurtado and Poole 2022).

In language learning, language cannot be considered in isolation from the context and the changing factors, such as sustainable development, that influence it (Paesani and Allen 2012). Current language teaching and learning, also called post-method or post-communicative pedagogy (Melin 2019), does not promote a single dominant method of teaching, and language teachers use a variety of methods influenced by the teaching context, which includes their institution and the learners (Kumaravadivelu 2006; Wesely et al. 2021). The modern approach to language teaching and learning views language as dynamic (Larsen-Freeman 2019) and shared, both culturally and historically, among all language users (e.g. Cowley 2011; Dufva et al. 2014). In transformative language teaching that aims at advancing sustainability, the principles of transformation-oriented ESD are integrated into language teaching primarily with pedagogical methods that have been found to foster critical thinking, creativity, and collaboration (Brundiers et al. 2021; Rieckmann 2018) and advancing the SDGs by developing learners' sustainability competencies, such as systems thinking, anticipatory, normative, strategic, and interpersonal competencies (Redman and Wiek 2021). Language is also considered to be vital in advancing democratic goals, such as active citizenship, and the development of critical thinking and dialogue skills (Barili and Byram 2021; Byram 2014). These goals can be achieved, for instance, through group work that involves students from different countries working together on topics of social importance and using the target language for interaction. These activities enable students to analyse and reflect on issues from an international perspective and take civic action in their own communities to bring about positive democratic change (Porto et al. 2018).

This volume aims to help language educators integrate sustainability education, such as ESD, into language teaching practices by presenting theoretical knowledge and case studies from a variety of educational contexts. In this introductory chapter, we first define the relevance of a transformative approach to language teaching following the conceptualisation of Maijala et al. (2024). We then discuss the importance of sustainability in language teaching by reviewing how previous studies have implemented ESD in language teaching contexts. Next, we discuss the possibilities and challenges of implementing transformation-oriented ESD in language teaching and language teacher education. Finally, we briefly present the articles in this volume.

Disciplinary Framing of Transformative Orientation in Language Education

During the last few decades, language teaching has shifted towards a more collective approach, with a greater emphasis on the active participation of the learner (e.g. Dörnyei and Kormos 2000; Hiver et al. 2021). In addition, Kramsch's (2002) ecological perspective, which suggests that physical, social, and other environmental factors shape interaction, brought a sustainability education view to language learning research. Kramsch's (2009, 2021) contributions to language education, which emphasise linguistic diversity and the power of language, are in accordance with the principles of sustainability education. Furthermore, new openings in the field of language education contribute to SDGs and the promotion of ESD. For example, cultural humility is a concept primarily associated with the field of psychology that can be used to promote ESD in language education (Heikkola et al. forthcoming). Cultural humility surpasses cultural competence by highlighting the importance of self-reflection and understanding one's own beliefs and cultural diversity in order to work effectively with individuals from diverse cultural backgrounds (Fisher 2020). These skills are essential when promoting ESD in language education. The changes in methods and approaches by modern theories in the field of language education have made it possible to integrate SDGs into language teaching using holistic, learner-oriented methods commonly used in teaching sustainability competencies, such as enquiry-based learning (e.g. Alcantud-Díaz and LLoret-Catalá 2023).

Although there is consensus among scholars that the field of language education has entered a post-method era in which there is no longer a dominant method or paradigm of teaching languages, but rather a diversity of methods (Kumaravadivelu 2006; Wesely et al. 2021), we now take a brief look into the disciplinary traditions in language education. Traditionally, presentations of different methodologies in language teaching begin with the concept of *communicative competence*, first introduced by Hymes in the 1970s. It shifted the focus to learning through participation in meaningful communicative situations and encouraged the use of the target language in everyday situations. The current communicative approaches in language teaching are based on sociocultural theory (Lantolf and Thorne 2006). Accordingly, language learning can be seen as an ongoing reciprocal process between individuals and the world around them (van Lier 2008), relying on human contact and interaction. As communication is the central goal in language teaching, it is often associated with SDGs in terms of democracy (Byram 2014) and multilingualism (Rizvi and Lingard 2010; Stein-Smith 2016).

In general, ESD promotes learner-centred teaching approaches (e.g. project-based learning and task-based learning) that enhance the learning process (Sinakou et al. 2019). In recent decades, learner activity and engagement have become important aspects of language learning (e.g. Dörnyei and Kormos 2000; Hiver et al. 2021). The modern approach to language teaching is dynamic and complex and can be regarded as opposite to the foundations of the Common European Framework of Reference for Languages (CEFR; Council of Europe 2001, 2020), which are based on the linear and traditional school-based mode of language learning. The concept of language as dynamic has been further elaborated recently in the development of dynamic systems theory (DST), a theory adopted from the natural sciences by Larsen-Freeman (2019). DST views language as a complex, non-linear, fluid, and self-organising element that is open to feedback and capable of self-adaptation. This perspective on language as a phenomenon has implications for language learning and teaching. Language learning is a dynamic process, and language itself is constantly changing (Larsen-Freeman 2019). In their edited book, Leaver (2021) and other authors lean on Mezirow's (1978) transformation theory when they posit that transformative language learning and teaching represents the next paradigm shift in the field of language teaching and learning. The transformative approach to teaching may result in learner change in various domains, including cognition, behaviour, values and emotion. The objective of this approach is for learners to develop critical and creative thinking through learner-centred activities, such as analysis, evaluation, exploration, comparison and research projects. Consequently, learners should experience a personal transformation.

Change or transformation is at the heart of SDGs, and teachers are important agents in this process (Bürgener and Barth 2018; Al Amin and Greenwood 2018; Weinberg et al. 2020; Koskela and Kärkkäinen 2021). The teacher's role is to catalyse change in their students through transformation-oriented pedagogy so that they contribute to achieving SDGs in their own lives. ESD, the concept developed by UNESCO (2018), is the key to achieving SDGs. Its core content is to introduce themes that focus on SDGs in all forms of learning (UNESCO 2018, 2021). Therefore, the aim of ESD is to prepare individuals with the necessary skills and knowledge to contribute to a more sustainable future by fostering critical thinking, creativity, and collaboration through experiential learning and community engagement (Brundiers et al. 2021; Rieckmann 2018). The holism-pluralism-action framework (Sinakou et al. 2019) forms the basis of the transformation-oriented ESD approach, which is based on pedagogical concepts and learning processes that support change. The review article by Redman and Wiek (2021) identified a number of key competences for educating students to promote change towards sustainable development, including systems thinking, anticipatory, normative, strategic and interpersonal competencies. Systems thinking and interpersonal competences have been most widely used in language teaching, for instance, engaging students in solving real-world challenges through project-based learning (Konrad et al. 2020) or incorporating multilingual and multicultural dimensions of climate change into language courses (Ferry 2021).

Transformative orientation is needed not only in subjects such as biology or geography but also in language education. Language teaching has the potential to implement ESD because the pedagogical processes of contemporary language teaching and ESD are commensurate with one another. This potential has been recognised in research (Arslan and Curle 2021; de la Fuente 2021; Kwee 2021; Maijala et al. 2023a; Tochon 2009), but it has not yet been exploited. Many approaches in language education can contribute to ESD. For instance, linguistically responsive teaching (i.e. supporting learners from diverse backgrounds; Lucas and Villegas 2013) enhances equity and social justice (Maijala et al. 2023a). Also, learner-oriented teaching (de la Fuente 2021), project-based learning (Ferry 2021) and content and language integrated learning (CLIL) (Barbas-Rhoden 2021) can provide language learners with transferable skills, including problem-solving, conflict resolution, and the ability to act responsively in society. These skills, enhancing change agency, are the basis for many competencies needed to equip learners with sustainability and change agency skills (Boeve-de Pauw et al. 2015; Frisk and Larson 2011). The model of *Transformative Language Teaching for Sustainability* (TLS; Maijala et al. 2024) provides a novel and useful framework for language education and fills a gap in the field of language education for accessible theory to help language teachers put ESD into practice at all levels of education. The model aims to support language teachers in identifying topics and using methods that promote ESD and thus integrate it into language teaching. At the content level, the TLS model moves from individual language learning purposes towards more accessible and global goals, including topics such as the environment and climate change, and activities to enhance both individual and community responsibility for the environment. The model introduces learner-centred activities, such as discussions and problem-solving tasks, which are highlighted both in the CEFR (Council of Europe 2001, 2020) and in the research literature on sustainability education (e.g. Sinakou et al. 2019).

Significance of ESD in Language Teaching

Although there is a need to focus more research on the implementation of ESD in language teaching (Maijala et al. 2024), we want to point out that there is previous research available, especially on ecological considerations in the field of applied linguistics (e.g. Haugen 1971; for a review, see Do Couto 2014; Katunich and Goulah 2020; Molina 2022). Among others, the posthumanist approach situates language within the context of human interaction, the physical objects and nature around us (Pennycook 2018) and ecolinguistics examines 'the role of language in the life-sustaining interactions of humans, other species and the physical environment' (Stibbe 2021, p. 223). Linking climate change to English language teaching (Goulah 2017) and Stibbe's (2021) contributions to practical pedagogical solutions on ecolinguistics build a link to language education by, for instance, providing an example of how the mediation of ecolinguistics can happen in classrooms or how ecological texts and language use can be explored by learners. Nevertheless, it appears that the ecolinguistic approach has not yet been fully integrated into the wider field of language teaching and learning.

In European language teaching, the CEFR (Council of Europe 2001, 2020) provides guidelines for planning curricula and textbooks and, consequently, for choosing topics in language teaching (Byram 2022). The CEFR emphasises communicative activities and the action-oriented approach, which are closely linked with the principles of communicative language teaching. The framework focuses on practical language use and communication. The updated framework, the CEFR Companion volume, also emphasises intercultural aspects such as mediation, multilingualism, and cooperative learning (Council of Europe 2020), which can be seen as essential when trying to achieve SDGs and promote ESD. However, the CEFR does not directly mention SDGs or how they should be included in language lessons.

Recently, the significance of integrating sustainability in language teaching has emerged in studies from the US, showcasing examples of different approaches to implementing ESD in language teaching and learning. The volume edited by Melin (2019) provides examples of projects in the area of environmental humanities that combine language teaching and environmental issues. Among examples were curricular projects focusing on literature teaching in French studies (Carrico 2019) and on designing an oral communication course in Spanish on environmental and social sustainability in Latin America (de la Fuente 2019). In addition, the volume presents pedagogical innovations which include sustainability issues in content-based language courses by using authentic teaching materials in the target language: The use of authentic multimedia material on the internet (e.g. Barnes-Karol and Broner 2010), such as visual images and blogs, can raise learners' awareness of different cultural perspectives and environmental attitudes.

Another recently published volume from the US by de la Fuente (2021) contains case studies of ESD implementation in colleges across the US, for example, by Ferry (2021) through project-based learning and by Barbas-Rhoden (2021) using the CLIL approach, and by Reisinger et al. (2021), who integrated learning Chinese, French and Spanish in an Environmental Science course. The last case study example described how project-based enquiry and community-engaged learning were used as teaching methods and the materials were related to environmental 'wicked problems', such as environmental justice, climate change, hydropolitics, and biodiversity in the learners' country of origin. These examples from the US show that sustainability-related teaching projects enable students to make connections between sustainability topics and the study of language. For instance, students' perceptions about linking sustainability and language studies were mostly positive and students' language proficiency developed along with their understanding of environmental studies (e.g. Reisinger et al. 2021).

Overall, the integration of sustainable development in language teaching seems to have a positive influence on both students' language learning processes and their environmental awareness (e.g. Hossain 2018; Jodoin 2020) and some research indicates that language teachers are already implementing ESD in their teaching (Goulah and Katunich 2020). Thus, it seems that integrating sustainability topics and methods does not disturb the language learning process but gives it purpose and integrates it into a wider holistic context.

Challenges and Possibilities Associated with Implementing ESD

In general, the research literature reveals challenges in implementing ESD in practice, regardless of the subject. For instance, little research is available on the practical implementation of SDGs in language education (Astawa et al. 2024). The difficulties encountered in implementing ESD can be linked to the interpretation of sustainable development, which is influenced by local, regional, and global contexts (Bonnett 2002). Moreover, sustainable development issues are broad and multidimensional, often requiring a multidisciplinary approach (Annan-Diab and Molinari 2017). Disciplinary boundaries can sometimes be an obstacle to multidisciplinary thinking related to sustainability, not only at universities (Abbonizio and Ho 2020) but also in schools (Sund and Gericke 2020). In school contexts, teachers have expressed several barriers and obstacles to teaching sustainability, such as the lack of teaching materials, training, and knowledge of how to put ESD into practice (e.g. Waltner et al. 2020; Letouzey-Pasquier et al. 2023).

There is increasing pressure worldwide to integrate ESD into language teaching (e.g. Katunich and Goulah 2020). However, language education faces specific challenges because, unlike other subjects, language is both the medium and the content of teaching (Hammadou and Bernhardt 1987; Halliday 1993). In addition, most language teachers apply no ESD in their teaching (Borg et al. 2012, 2014; McClanahan 2013; Sinakou et al. 2017; Sund et al. 2020), possibly because of the characteristics of language teaching that are different from other subjects (Borg 2006) or disciplinary traditions (e.g. Sund and Gericke 2021). While some language teachers teach aspects of sustainability without recognising their role in sustainable development (Laine et al. 2023), many language teachers who apply ESD do seem to be personally committed to the goals of sustainable development (Fischer and Hänze 2020; Gustafsson et al. 2015; Kuusalu et al., forthcoming). Teachers' familiarity with teaching practices, understanding of their readiness to teach sustainability phenomena, and motivation and attitudes influence how extensively they emphasise different dimensions of sustainable development and their implementation of ESD (Kwee 2021; Maijala et al. 2023b; Wiek et al. 2015; of agency in language teaching see also Kuusalu et al., forthcoming). Consequently, teachers' pedagogical choices influence learners' experiences with sustainable development (Öhman and Sund 2021). The use of motivating teaching methods is likely to correlate positively with learners' perceptions of sustainability and its significance (Frisk and Larson 2011; Jodoin 2020), and to promote learners' change agency (Huss-Lederman 2020). Earlier research suggests that multiple holistic teaching approaches and systems thinking are needed to incorporate ESD into language teaching (Kuusalu et al. 2024); this includes a pluralistic approach to language teaching that centres on the concepts of activity, dialogue, and diversity, and allows learners to express their views through discussions, thereby promoting the ESD goals. The need to include holistic and learner-centred ESD practices in language teaching leads to an essential question: How can language teachers be supported in the implementation of transformation-oriented ESD?

This Volume

The aim of this edited volume is to assist language teaching practitioners in their integration of sustainability education, for example in the form of ESD, into language education. It draws on research and case studies from a variety of educational contexts and makes Maijala et al.'s (2024) TLS model accessible for use in language teaching practice. There is often a discrepancy between top-down decisions to integrate sustainability education into language teaching and the actual implementation of these decisions. Studies indicate that both in-service and pre-service language teachers lack the requisite qualifications to incorporate ESD into their language courses. This edited volume aims to address this potential shortcoming and to support language teachers and language teacher educators at all educational levels who seek to engage with ESD. The authors are experienced scholars or teacher-scholars active in different areas of foreign language education and science education and thus represent a variety of educational contexts. They all have experience dealing with sustainability-related topics in their research and classroom practices. Due to the novelty of the transformative approach, there are few contributors in the field of language teaching and learning. This explains why most of the studies in this volume have been conducted as collaborative research, with contributions from the editors. Although the articles are examples of work in specific languages, the outcomes can be adapted to other languages, subjects, and educational contexts. The chapters of this volume can be read separately, in any order, or as a whole.

Each chapter introduces the topic and explains its relevance in a broader context. Then, practical recommendations, teaching principles, and applications for language educators are presented. The aim of each chapter is to discuss the possibilities and challenges of implementing sustainability education, such as transformation-oriented ESD, in language teaching and language teacher education in different educational settings. The aim is to provide language educators with the tools to transform theoretical knowledge into practical skills. To this end, the volume has been divided into two parts: Theoretical considerations and Pedagogy and practices.

In the first part, the 'Theoretical considerations', three chapters approach ESD from different perspectives. In Chapter 2, Niklas Gericke elaborates on the political background of ESD and the challenges this dimension brings to educators. Using the concept of powerful knowledge, he discusses the problems of balance between specialised knowledge stemming from disciplinary traditions, which is essential when dealing with complex problems, and ESD policy, which has the potential to increase the relevance of education by connecting what is taught with real-life issues. Gericke uses a multidisciplinary perspective and tightly links language education to the wider field of teaching ESD.

In Chapter 3, Salla-Riikka Kuusalu presents aspects of ESD, such as its learning processes, which have been shown in existing research to be supportive of language learning, and discusses how ESD has developed and could evolve to further enhance learning and learner agency and catalyse transformative action. Kuusalu considers ESD from the learner's perspective by reviewing relevant research literature and focusing on transformative ESD methods and practices both in the Western world and in the context of the Global South. Kuusalu begins by highlighting the modern view of language as a complex adaptive system and the effect that this view has on learner agency. Next, the learning processes of ESD are considered from the viewpoint of language learning, and the discussion progresses to transferrable sustainability competencies acquired through language education and their relevance to solving 'wicked problems'. Finally, Kuusalu highlights the importance of increasing both the critical literacy of sustainable development concepts and epistemic justice, particularly in language education.

Chapter 4 by Attila Krizsán discusses the potential of linguistic research in aiding the move towards a more sustainable global ecosystem. Krizsán argues that linguistic research—particularly functional-, critical-

and ecolinguistics—could contribute to reconceptualizing and changing sustainability-related communication, and, through this, thinking and action. Krizsán postulates that these changes could give ways to new, more sustainable educational and cultural practices and lead to positive alterations of human–nature relationships.

The chapters in part two, 'Pedagogy and practices', present practical studies showing various potential methods and approaches for transformative ESD in language classes. Chapter 5 by a collaborative research group (Salla-Riikka Kuusalu, Minna Maijala, Meameno Aileen Shiweda, Minna Maunula, Minna Maunumäki, and Soila Merijärvi) analyses the reflections on sustainable development of Namibian and Finnish pre-service language teachers in a comparative study. The study focuses on how the students envision the future role of sustainable development in language teaching and what pedagogical practices they consider sustainable. What becomes evident in the analysis is that the focus when thinking about sustainability is often context dependent and that pre-service language teachers' future visions vary between different contexts. Notably, the Finnish students tended to focus on local actions and shied away from global issues, whereas the Namibian students linked the learning outcomes more clearly with society, called for teacher's responsibility in leading sustainability transitions, and highlighted continuous education in increasing teacher's sustainability competence. The authors share their views on the similarities and differences in the results and suggest ways to incorporate SDGs into language lessons.

Textbook analysis is the core of Chapter 6, where Minna Maijala, Mareen Patzelt, and Friederike Delius investigate the sustainability content of German textbooks for beginners used in five European countries. The authors discuss the vital role textbooks have in language learning and teaching and why their content should be developed to include ESD materials and teaching methods. The role of language teachers in facilitating discussions about sustainability is also considered. Marianne Broermann, Salla-Riikka Kuusalu and Minna Maijala conducted research on transformative language teaching practices in the context of higher education in Germany. In Chapter 7, they present a study in which adult L2 learners' conceptions of sustainability were examined before and after attending two Swedish language classes that used a case study method tailored for the study. The findings show that using concrete, real-life cases in language teaching allows learners to more profoundly understand complex issues related to sustainability, such as Indigenous rights in the transition towards renewable energy and the challenges tourism poses to Sámi people in their homelands.

In Chapter 8, Minna Maijala, Mareen Patzelt, Päivi Laine, and Soila Merijärvi describe in detail the process of developing engaging language teaching activities for different learner groups and how these were experimented in two language teaching contexts in higher education. The materials covering all dimensions of sustainable development (ecological, social, cultural and economic) are based on Maijala et al.'s (2024) TLS model and on the latest research literature in the fields of ESD and language education. The chapter discusses how language teachers can adapt the content of language textbooks and other materials and implement SDGs in their lessons.

In Chapter 9, Leena Maria Heikkola, Minna Maijala, Maarit Mutta, and Päivi Laine introduce a research studying pre-service language teachers' perceptions of multilingualism. Their analysis was based on the responses to an online survey which focused on the definitions of multilingualism, actions, and use of language in relation to multilingualism. The research group found, for example, that multilingualism was mostly perceived to be a characteristic of an individual, and that the students acknowledged multiple ways how linguistic repertoires work together and contexts, where multilingualism becomes evident or is useful. Continuing with the theme of multilingualism, Santiago Sánchez Moreano draws attention to the plurilingual repertoires of pupils in a multi-ethnic school in Leticia in Columbian Amazonia. Chapter 10 describes how languages can be used in teaching and learning to create spaces for both social and epistemic justice. His research on the sixteen minority groups in the area increases knowledge of the possibilities to promote Indigenous knowledge and empowerment in multi-ethnic schools through pupils' own language repertoires.

The book concludes with Chapter 11, where Salla-Riikka Kuusalu and Arran Stibbe introduce ecolinguistic learning modules as a tool for teaching about sustainability especially in the context of English language and linguistics. The authors challenge anthropocentrism in sustainability and its education and propose an alternative narrative to replace the current, unsustainable way of thinking and living. By bringing together recent research, this volume starts a new dialogue on opportunities to integrate sustainability education, such as ESD, into language education. This volume presents not only research outcomes but also practical recommendations, teaching principles, and applications for language educators willing to implement the insights of research.

References

Abbonizio, J.K., and S.S.Y. Ho. 2020. Students' perceptions of interdisciplinary coursework: An Australian case study of the master of environment and sustainability. *Sustainability* 12 (21) (October 27): 8898. https://doi.org/10.3390/su12218898.

Alcantud-Díaz, M., and C. Loret-Catalá. 2023. Bridging the gap between teacher training and society. Sustainable Development Goals (SDGs) in English as a foreign language (EFL). *Globalisation, Societies and Education* (November 20): 1–12. https://doi.org/10.1080/14767724.2023.2283510.

Al Amin, M.A., and J. Greenwood. 2018. The UN Sustainable Development Goals and teacher development for effective English teaching in Bangladesh: A gap that needs bridging. *Journal of Teacher Education for Sustainability* 20 (2) (December 1): 118–138. https://doi.org/10.2478/jtes-2018-0019.

Annan-Diab, F., and C. Molinari. 2017. Interdisciplinarity: Practical approach to advancing education for sustainability and for the Sustainable Development Goals. *The International Journal of Management Education* 15 (2) (July): 73–83. https://doi.org/10.1016/j.ijme.2017.03.006.

Arslan, S., and S. Curle. 2021. Sustainable Development Goals in the English language high school curriculum in Turkey. *European Journal of Education* 56 (4) (December): 681–695. https://doi.org/10.1111/ejed.12473.

Astawa, N.L.P.N.S.P., M.H. Santosa, L.P. Artini, and P.K. Nitiasih. 2024. Sustainable Development Goals in EFL students' learning: A systematic review. *Pegem Journal of Education and Instruction* 14 (2): 1–11. https://doi.org/10.47750/pegegog.14.02.01.

Barbas-Rhoden, L. 2021. Sustainability and the pluriverse: From environmental humanities theory to content-based instruction in Spanish curricula.

In *Education for Sustainable Development in foreign language learning*, ed. M.J. de la Fuente, 17–33. New York: Routledge.

Barili, A., and M. Byram. 2021. Teaching intercultural citizenship through intercultural service learning in world language education. *Foreign Language Annals* 54 (3) (October): 776–799. https://doi.org/10.1111/flan.12526.

Barnes-Karol, G., and M.A. Broner. 2010. Using images as springboards to teach cultural perspectives in light of the ideals of the MLA report. *Foreign Language Annals* 43 (3) (September): 422–445. https://doi.org/10.1111/j.1944-9720.2010.01091.x.

Boeve-de Pauw, J., N. Gericke, D. Olsson, and T. Berglund. 2015. The effectiveness of Education for Sustainable Development. *Sustainability* 7 (11) (November 24): 15693–15717. https://doi.org/10.3390/su71115693.

Bonnett, M. 2002. Education for sustainability as a frame of mind. *Environmental Education Research* 8 (1) (February): 9–20. https://doi.org/10.1080/13504620120109619.

Borg, S. 2006. The distinctive characteristics of foreign language teachers. *Language Teaching Research* 10 (1) (January): 3–31. https://doi.org/10.1191/1362168806lr182oa.

Borg, C., N. Gericke, H.-O. Höglund, and E. Bergman. 2012. The barriers encountered by teachers implementing Education for Sustainable Development: Discipline bound differences and teaching traditions. *Research in Science & Technological Education* 30 (2) (July): 185–207. https://doi.org/10.1080/02635143.2012.699891.

Borg, C., N. Gericke, H.-O. Höglund, and E. Bergman. 2014. Subject- and experience-bound differences in teachers' conceptual understanding of sustainable development. *Environmental Education Research* 20 (4) (July 4): 526–551. https://doi.org/10.1080/13504622.2013.833584.

Brundiers, K., M. Barth, G. Cebrián, M. Cohen, L. Diaz, S. Doucette-Remington, W. Dripps, G. Habron, K. Harré, M. Jarchow, K. Losch, J. Michel, Y. Mochizuki, M. Rieckmann, R. Parnell, P. Walker, and M. Zint. 2021. Key competencies in sustainability in higher education—Toward an agreed-upon reference framework. *Sustainability Science* 16 (1) (January): 13–29. https://doi.org/10.1007/s11625-020-00838-2.

Bürgener, L., and M. Barth. 2018. Sustainability competencies in teacher education: Making teacher education count in everyday school practice. *Journal of Cleaner Production* 174 (February): 821–826. https://doi.org/10.1016/j.jclepro.2017.10.263.

Byram, M. 2014. Twenty-five years on—from cultural studies to intercultural citizenship. *Language, Culture and Curriculum* 27 (3) (September 2): 209–225. https://doi.org/10.1080/07908318.2014.974329.

Byram, M. 2022. Politics, origins and futures of the CEFR. *The Language Learning Journal* 50 (5) (September 3): 586–599. https://doi.org/10.1080/09571736.2020.1845392.

Carrico, A. 2019. Reflections on water: Inspiring environmental consciousness through engagement with French texts. In *Foreign language teaching and the environment: Theory, curricula, institutional structures*, ed. C.A. Melin, 81–93. New York: Modern Language Association.

Council of Europe. 2001. *Common European framework of reference for languages: Learning, teaching, assessment*. Strasbourg: Council of Europe. Available from: https://rm.coe.int/1680459f97 (accessed August 6, 2024).

Council of Europe. 2020. *Common European framework of reference for languages: Learning, teaching, assessment—Companion volume*. Council of Europe Publishing, Strasbourg. Available from: www.coe.int/lang-cefr (accessed July 16, 2024).

Cowley, S.J. 2011. Distributed language. In *Distributed language*, ed. S.J. Cowley, 1–14. Amsterdam: John Benjamins.

de la Fuente, M.J. 2019. Stepping out of the language box: College Spanish and sustainability. In *Foreign language teaching and the environment: Theory, curricula, institutional structures*, ed. C.A. Melin, 130–145. New York: The Modern Language Association.

de la Fuente, M.J. 2021. *Education for Sustainable Development in foreign language learning: Content-based instruction in college-level curricula*, 1st ed. New York: Routledge.

Do Couto, H.H. 2014. Ecological approaches in linguistics: A historical overview. *Language Sciences* 41 (January): 122–128. https://doi.org/10.1016/j.langsci.2013.08.001.

Dufva, H., M. Aro, and M. Suni. 2014. Language learning as appropriation: How linguistic resources are recycled and regenerated. *AFinLA-e* 6 (June): 20–31. Available from: https://journal.fi/afinla/article/view/46278 (accessed November 14, 2024).

Dörnyei, Z., and J. Kormos. 2000. The role of individual and social variables in oral task performance. *Language Teaching Research* 4 (3) (July): 275–300. https://doi.org/10.1177/136216880000400305.

Eilam, E., and T. Trop. 2010. ESD pedagogy: A guide for the perplexed. *The Journal of Environmental Education* 42 (1) (September 17): 43–64. https://doi.org/10.1080/00958961003674665.

Fazio, X., and D.D. Karrow. 2013. Negotiating the constraints of schools: Environmental education practices within a school district. *Environmental Education Research* 19 (5) (October): 639–655. https://doi.org/10.1080/13504622.2012.729812.

Ferry, M.M. 2021. Project-based language learning: Addressing cultural and linguistic diversity issues in climate action. In *Education for Sustainable Development in foreign language learning: Content-based instruction in college-level curricula*, ed. M.J. de la Fuente, 1st ed., 50–64. New York: Routledge.

Fischer, E., and M. Hänze. 2020. How do university teachers' values and beliefs affect their teaching? *Educational Psychology* 40 (3) (March 15): 296–317. https://doi.org/10.1080/01443410.2019.1675867.

Fisher, E.S. 2020. Cultural humility as a form of social justice: Promising practices for global school psychology training. *School Psychology International* 41 (1) (February): 53–66. https://doi.org/10.1177/0143034319893097.

Frisk, E., and K.L. Larson. 2011. Educating for sustainability: Competencies & practices for transformative action. *Journal of Sustainability Education* 2 (1) (March): 1–20.

Goulah, J. 2017. Climate change and TESOL: Language, literacies, and the creation of eco-ethical consciousness. *Tesol Quarterly* 51 (1): 90–114.

Goulah, J., and J. Katunich, ed. 2020. *TESOL and sustainability: English language teaching in the Anthropocene era*. Bloomsbury Advances in Ecolinguistics. London New York Oxford New Delhi Sydney: Bloomsbury Academic.

Gustafsson, P., S. Engström, and A. Svenson. 2015. Teachers' view of sustainable development in Swedish upper secondary school. *Procedia—Social and Behavioral Sciences* 167 (January): 7–14. https://doi.org/10.1016/j.sbspro.2014.12.635.

Halliday, M.A.K. 1993. Towards a language-based theory of learning. *Linguistics and Education* 5: 93–116.

Hammadou, J., and E.B. Bernhardt. 1987. On being and becoming a foreign language teacher. *Theory Into Practice* 26 (4) (September): 301–306. https://doi.org/10.1080/00405848709543290

Haugen, E. 1971. The ecology of language. *Linguist* 13 (25): 19–26.

Heikkola, L.M., M. Maijala, S. Merijärvi, P. Laine, M. Mutta, forthcoming. Finnish pre-service language teachers' understanding of culture—from cultural humility to social justice. *Manuscript available upon request from the authors.*

Hiver, P., A.H. Al-Hoorie, J.P. Vitta, and J. Wu. 2021. Engagement in language learning: A systematic review of 20 years of research methods and definitions. *Language Teaching Research* (March 24): 136216882110012. https://doi.org/10.1177/13621688211001289

Holden, E., K. Linnerud, and D. Banister. 2014. Sustainable development: Our common future revisited. *Global Environmental Change* 26 (May): 130–139. https://doi.org/10.1016/j.gloenvcha.2014.04.006.

Hossain, M. 2018. Enacting green pedagogy in the EFL classrooms in Bangladesh: Prospects, challenges and pragmatic breakthroughs. *Language Education Forum* 1 (1): 1–5. Available from: http://pubs.sciepub.com/lef/1/1/1 (accessed January 6, 2024).

Huss-Lederman, S. 2020. Investigating environmental sustainability in an English writing course for international students: PBL as an on-ramp to academic belonging. In *Project-based learning in second language acquisition: Building communities of practice in higher education*, 286. Oxon: Routledge.

Jodoin, J.J. 2020. Promoting language Education for Sustainable Development: A program effects case study in Japanese higher education. *International Journal of Sustainability in Higher Education* 21 (4) (May 14): 779–798. https://doi.org/10.1108/IJSHE-09-2019-0258.

Katunich, J., and J. Goulah. 2020. Introduction: TESOL and sustainability. In *TESOL and sustainability: English language teaching in the Anthropocene era*, ed. J. Goulah, and J. Katunich, 1st ed. London: Bloomsbury Publishing.

Konrad, T., A. Wiek, and M. Barth. 2020. Embracing conflicts for interpersonal competence development in project-based sustainability courses. *International Journal of Sustainability in Higher Education* 21 (1) (January 6): 76–96. https://doi.org/10.1108/IJSHE-06-2019-0190.

Koskela, T., and S. Kärkkäinen. 2021. Student teachers' change agency in Education for Sustainable Development. *Journal of Teacher Education for Sustainability* 23 (1) (June 1): 84–98. https://doi.org/10.2478/jtes-2021-0007.

Kramsch, C., ed. 2002. *Language acquisition and language socialization: Ecological perspectives*. London: Continuum.

Kramsch, C. 2009. *The multilingual subject: What foreign language learners say about their experience and why it matters*. Oxford: Oxford University Press.

Kramsch, C. 2021. *Language as symbolic power*. Cambridge: Cambridge University Press.

Kumaravadivelu, B. 2006. *Understanding language teaching: From method to postmethod*. London: Routledge.

Kuusalu, S.-R., M. Patzelt, and M. Maijala., forthcoming. Language teachers as educators for sustainability in Finland—Factors influencing the experience of agency. *Manuscript available upon request from the authors.*

Kuusalu, S.-R., P. Laine, M. Maijala, M. Mutta, and M. Patzelt. 2024. University language students' evaluations of ecological, social, cultural and economic sustainability and their importance in language teaching. *International Journal of Sustainability in Higher Education* 25 (9) (January 1): 1–18. https://doi.org/10.1108/IJSHE-05-2023-0169.

Kwee, C.T.T. 2021. I want to teach sustainable development in my English classroom: A case study of incorporating Sustainable Development Goals in English teaching. *Sustainability* 13 (8) (April 9): 4195. https://doi.org/10.3390/su13084195

Laine, P., M. Maijala, and K. Mäntylä. 2023. Kestävän kehityksen periaatteet kielenoppimisen arvioinnissa: The principles of sustainable development in assessing language learning. In *Kielenoppimista tukeva arviointi—Assessment for supporting language learning*, ed. T. Mäkipää, R. Hilden, and A. Huhta. *A FinLA-e* 15, 162–183. https://doi.org/10.30660/afinla.124924.

Lantolf, J.P., and S.L. Thorne. 2006. *Sociocultural theory and the genesis of second language development*. Oxford University Press.

Larsen-Freeman, D. 2019. On Language learner agency: A complex dynamic systems theory perspective. *The Modern Language Journal* 103 (January): 61–79. https://doi.org/10.1111/modl.12536.

Leaver, B.L. 2021. Transformative language learning and teaching: The next paradigm shift and its historical context. In *Transformative language learning and teaching*, ed. B.L. Leaver, D.E. Davidson, and C. Campbell, 1st ed. Cambridge University Press. Available from: https://www.cambridge.org/core/product/identifier/9781108870788/type/book. (accessed October 3, 2024).

Letouzey-Pasquier, J., B. Gremaud, S. Blondin, and P. Roy. 2023. Development of teachers' practices in the field of Education for Sustainable Development (ESD): A discursive community of interdisciplinary practices focusing on the theme of chocolate. *Environmental Education Research* 29 (8) (August 3): 1155–1169. https://doi.org/10.1080/13504622.2022.2128056.

Lewis, E., C. Mansfield, and C. Baudains. 2014. Ten tonne plan: Education for sustainability from a whole systems thinking perspective. *Applied Environmental Education & Communication* 13 (2) (April 3): 128–141. https://doi.org/10.1080/1533015X.2014.950890.

Lozano, R., M. Merrill, K. Sammalisto, K. Ceulemans, and F. Lozano. 2017. Connecting competences and pedagogical approaches for sustainable development in higher education: A literature review and framework proposal. *Sustainability* 9 (10) (October 20): 1889. https://doi.org/10.3390/su9101889.

Lucas, T., and A.M. Villegas. 2013. Preparing linguistically responsive teachers: Laying the foundation in preservice teacher education. *Theory into Practice* 52 (2) (April): 98–109. https://doi.org/10.1080/00405841.2013.770327.

Maijala, M., L.M. Heikkola, S.-R. Kuusalu, P. Laine, M. Mutta, and K. Mäntylä. 2023a. Pre-service language teachers' perceptions of sustainability and its implementation in language teaching. *Language Teaching Research* (May 13): 136216882311706. https://doi.org/10.1177/13621688231170682.

Maijala M., P. Laine, M. Mutta, and S.-R. Kuusalu. 2023b. Kestävä kehitys yliopistojen kieltenopetuksessa [Sustainable language teaching at university]. Yliopistopedagogiikka 2. Available from: https://lehti.yliopistopedagogiikka.fi/2023/12/22/kestava-kehitys-yliopistojen-kieltenopetuksessa/ (accessed July 22, 2024).

Maijala, M., N. Gericke, S.-R. Kuusalu, L.M. Heikkola, M. Mutta, K. Mäntylä, and J. Rose. 2024. Conceptualising transformative language teaching for sustainability and why it is needed. *Environmental Education Research* 1–20. https://doi.org/10.1080/13504622.2023.2167941.

McClanahan, L.G. 2013. The Greening of the language arts: Considering sustainability outside of the science classroom. *Journal of Sustainability Education* 4 (January): n.p.

Melin, C., ed. 2019. *Foreign language teaching and the environment: Theory, curricula, institutional structures.* Teaching languages, literature, and culture, 6. New York: The Modern Language Association of America.

Mezirow, J. 1978. Perspective transformation. *Adult Education* 28 (2) (January): 100–110. https://doi.org/.1177/074171367802800202.

Micalay-Hurtado, M.A., and R. Poole. 2022. Eco-critical language awareness for English language teaching (ELT): Promoting justice, wellbeing, and sustainability in the classroom. *Journal of World Languages* 8 (2) (October 3): 371–390. https://doi.org/10.1515/jwl-2022-0023.

Molina, P. 2022. Bibliographical review on sustainable literacy in language learning and teaching. *Language Learning in Higher Education* 12 (2) (October 26): 513–524. https://doi.org/10.1515/cercles-2022-2059.

Murillo-Vargas, G., C.H. Gonzalez-Campo, and D.I. Brath. 2020. Mapping the integration of the Sustainable Development Goals in universities: Is it a field of study? *Journal of Teacher Education for Sustainability* 22 (2) (December 1): 7–25. https://doi.org/10.2478/jtes-2020-0013.

Ntona, E., A. Georgopoulos, G. Malandrakis, and P. Ragkou. 2024. Teachers' barriers dealing with environmental education programs' implementation in Greek secondary schools. *Environmental Education Research* 30 (5) (May 3): 700–719. https://doi.org/10.1080/13504622.2023.2182257.

Öhman, J., and L. Sund. 2021. A didactic model of sustainability commitment. *Sustainability* 13 (6) (March 11): 3083. https://doi.org/10.3390/su13063083.

Paesani, K., and H.W. Allen. 2012. Beyond the language-content divide: Research on advanced foreign language instruction at the postsecondary level. *Foreign Language Annals* 45 (s1) (June). Available from: https://doi.org/10.1111/j.1944-9720.2012.01179.x. (accessed October 22, 2024).

Pennycook, A. 2018. Posthumanist applied linguistics. *Applied Linguistics* 39 (4): 445–461.

Porto, M., S.A. Houghton, and M. Byram. 2018. Intercultural citizenship in the (foreign) language classroom. *Language Teaching Research* 22 (5): 484–498. https://doi.org/10.1177/1362168817718580.

Redman, A., and A. Wiek. 2021. Competencies for advancing transformations towards sustainability. *Frontiers in Education* 6: 785163.

Reisinger, D., S. Valnes Quammen, Y. Liu, and E. Virguez. 2021. Sustainability across the curriculum: A multilingual and intercultural approach. In *Education for Sustainable Development in foreign language learning*, ed. M.J. de la Fuente, 197–214. Routledge.

Rieckmann M. 2018. Learning to transform the world: Key competencies in Education for Sustainable Development. In *Issues and trends in Education for Sustainable Development*, ed. A. Leicht, J. Heiss, and W.J. Byun, 39–60. UNESCO Publishing.

Rizvi, F., and B. Lingard. 2010. *Globalizing education policy*. London, New York: Routledge.

Sinakou, E., J. Boeve-de Pauw, and P. Van Petegem. 2017. Exploring the concept of sustainable development within Education for Sustainable Development: Implications for ESD research and practice. *Environment, Development and Sustainability* 21 (1): 1–10. https://doi.org/10.1007/s10668-017-0032-8.

Sinakou, E., V. Donche, J. Boeve-de Pauw, and P. Van Petegem. 2019. Designing powerful learning environments in Education for Sustainable Development: A conceptual framework. *Sustainability* 11 (21) (October 28): 5994. https://doi.org/10.3390/su11215994.

Stein-Smith, K. 2016. The role of multilingualism in effectively addressing global issues: The Sustainable Development Goals and beyond. *Theory and Practice in Language Studies* 6 (12) (December 1): 2254. https://doi.org/10.17507/tpls.0612.03.

Stibbe, A. 2021. *Ecolinguistics: Language, ecology and the stories we live by*, 2nd ed. London: Routledge.

Sund, P., and N. Gericke. 2020. Teaching contributions from secondary school subject areas to Education for Sustainable Development—A comparative study of science, social science and language teachers. *Environmental Education Research* 26 (6) (June 2): 772–794. https://doi.org/10.1080/13504622.2020.1754341.

Sund, P., and N. Gericke. 2021. More than two decades of research on selective traditions in environmental and sustainability education—Seven functions of the concept. *Sustainability* 13 (12) (June 8): 6524. https://doi.org/10.3390/su13126524.

Sund, P., N. Gericke, and G. Bladh. 2020. Educational content in cross-curricular ESE teaching and a model to discern teacher's teaching traditions. *Journal of Education for Sustainable Development* 14 (1): 78–97.

Tochon, F.V. 2009. The key to global understanding: World languages education—Why schools need to adapt. *Review of Educational Research* 79 (2) (June): 650–681. https://doi.org/10.3102/0034654308325898.

UNESCO. 2018. Global Action Programme on Education for Sustainable Development. Education for Sustainable Development and the SDGs. Policy brief: Advancing ESD policy. Available from: https://unesdoc.unesco.org/ark:/48223/pf0000246270 (accessed July 22, 2024).

UNESCO. 2021. Education for Sustainable Development lens: A policy and practice review tool. Available from: https://unesdoc.unesco.org/ark:/48223/pf0000190898 (accessed July 22, 2024).

van Lier, L. 2008. Ecological-semiotic perspectives on educational linguistics. In *The handbook of educational linguistics*, 597–605. Oxford, UK: Blackwell Publishing Ltd.

Wals, A.E.J. 2009. *Review of contexts and structures for Education for Sustainable Development: 2009*. France: UNESCO.

Waltner, E.-M., K. Scharenberg, C. Hörsch, and W. Rieß. 2020. What teachers think and know about Education for Sustainable Development and how they implement it in class. *Sustainability* 12 (4) (February 24): 1690. https://doi.org/10.3390/su12041690.

WCED. 1987. *Our common future*. World commission on environment and development: Oxford University Press, Oxford.

Weinberg, A.E., C.D. Trott, W. Wakefield, E.G. Merritt, and L. Archambault. 2020. Looking inward, outward, and forward: Exploring the process of transformative learning in teacher education for a sustainable future. *Sustainability Science* 15 (6) (November): 1767–1787. https://doi.org/10.1007/s11625-020-00831-9.

Wesely, P.M., R. Vyn, and D. Neubauer. 2021. Teacher beliefs about instructional approaches: interrogating the notion of teaching methods. *Language Teaching Research* (March 11): 136216882199218. https://doi.org/10.1177/1362168821992180.

Wiek, A., M. Bernstein, R. Foley, M. Cohen, N. Forrest, C. Kuzdas, B. Kay, and L. Withycombe Keeler. 2015. Operationalising competencies in higher Education for Sustainable Development. In *Handbook of higher Education for Sustainable Development*, ed. M. Barth, G. Michelsen, M. Rieckmann, and I. Thomas, 241–260. London: Routledge.

Zygmunt, T. 2016. Language Education for Sustainable Development. *Discourse and Communication for Sustainable Education* 7 (1) (December 1): 112–124. https://doi.org/10.1515/dcse-2016-0008.

Part I

Theoretical Considerations

2

Education for Sustainable Development in Subject-Specific Education and the Contribution of Language Teaching

Niklas Gericke

Introduction

Sustainability education, which is often referred to as Education for Sustainable Development (ESD) in policy contexts, is an educational approach based on the idea that education should promote the adoption of sustainable lifestyles among learners. In this chapter, the concepts of *sustainability education* and ESD are used interchangeably. There is a vast literature on the differences between these terms with respect to the feasibility of seeking to simultaneously achieve development and sustainability, and the potential incompatibilities of these two goals (e.g., Smyth 2006). These issues are not addressed here because both concepts in the context of education aim to transform the actions of the students in a more sustainable way.

N. Gericke (✉)
Department of Environmental and Life Sciences, Karlstad University, Karlstad, Sweden
e-mail: niklas.gericke@kau.se

M. Maijala et al. (eds.), *Transformative Language Teaching for Sustainability*,
https://doi.org/10.1007/978-3-031-85493-4_2

ESD has evolved as an educational approach in response to the advent of the Anthropocene, that is the period of time in which industrialisation and its aftermath have made humanity's impact on the environment strong enough to cause significant geological changes worldwide (Gericke 2022). Whereas conventional scholastic education emphasises the learning of content in the form of predefined knowledge or skills, ESD seeks to encourage students to act in more sustainable ways. This difference in aims raises a question: should sustainability education be implemented within the context of general or subject-specific education? Because ESD uses a competence-based approach, it is most commonly implemented within general education. However, based on Gert Biesta's (2010, 2017) suggestions about the main purposes of education and the work of Michael Young (2008, 2015) on the idea of powerful knowledge and possible trajectories of the curricula for the future, I will here argue for the value of a multidisciplinary perspective on sustainability education that relies on contributions from diverse subject-specific teaching traditions. Although previous research has shown that language teachers do not consider sustainability education central to their teaching, this chapter will draw on findings from empirical studies (Sund and Gericke 2020; Sund et al., 2020) to show how a transformative language education strategy can play a major role in empowering students' action competence for sustainability.

Education for Sustainable Development Addressing Complex and Wicked Problems

Sustainability education is a concept that has been developed at the policy level, often under the name of ESD, in response to the growing recognition that environmental education has limited potential to influence students' ways of living or to promote more sustainable lifestyles despite having been a major focus of the curriculum in many countries for more than 50 years (Gericke et al. 2020). Environmental education focuses on teaching students about specific environmental problems such as climate change, over-consumption, and ozone layer depletion. These topics are usually incorporated into the teaching of academic subjects that are directly relevant to key aspects of these problems, such as biology, geography, and civics. However, environmental education has also been provided as a subject in its own right—environmental science.

Environmental education focuses on giving students knowledge; regardless of the context in which the teaching occurs, the underlying objective is to transmit factual knowledge about how and why the problems emerge, and what could be done to counter them. It was implicitly assumed that a better understanding of the specific environmental problems would foster a kind of civic knowledge that would encourage people to act at personal and societal levels in ways that would gradually solve these problems (UNESCO 2017). Within this conceptual framework, environmental problems are seen from a disciplinary perspective (or from a subject perspective in schools) as specific problems that can be solved by improving the general knowledge level among students (and thereby, among citizens in general, as the students complete their education and move on in their careers).

Unfortunately, it has since become apparent that greater knowledge of environmental problems does not automatically translate into environmentally friendly behaviour (Kollmuss and Agyeman 2002). Therefore, environmental education and environmental education research have been reoriented to address affective aspects of teaching and learning in addition to cognitive aspects (Marcinkowski and Reid 2019), causing environmental education to be reoriented towards sustainability education, or ESD. This field has drawn inspiration as well as concepts relating to attitudes, values, emotions, and behaviour from psychological research, recognising that all aspects of students' consciousness must be addressed in order to effectively engender changes in the way they live and act.

Consequently, teaching approaches that seek to transform learners through education, which are often described as being more progressive than alternative approaches, have frequently been advocated in sustainability education research (Eilam and Trop 2010). These teaching approaches, which are positioned outside the syllabi of traditional school subjects, have collectively been named ESD. The need for sustainability education or ESD has been identified and propagated at the political level, often internationally within the United Nations and its subordinate bodies such as the United Nations Educational, Scientific and Cultural Organization (UNESCO). The ideas behind ESD have since emerged from reciprocal interactions between policy and educational research (Vare et al. 2022). The definition of ESD from UNESCOs educational policy is as follows:

> ESD does not only integrate contents such as climate change, poverty and sustainable consumption into the curriculum; it also creates interactive, learner-centred teaching and learning settings. What ESD requires is a shift from teaching to learning. It asks for an action-oriented, transformative pedagogy, which supports self-directed learning, participation and collaboration, problem-orientation, inter- and transdisciplinarity and the linking of formal and informal learning. Only such pedagogical approaches make possible the development of the key competencies needed for promoting sustainable development. (UNESCO 2017, p. 7)

The very first sentence of this passage refers to 'integrating contents' and thus calls for a multidisciplinary perspective on sustainability education. The definition also focuses on learning outcomes that are often related to specific content knowledge of the type conveyed within environmental education.

In ESD, the taught content is viewed holistically, and environmental problems are regarded as socio-political problems because they necessitate the resolution of socio-economic conflicts of interest (Öhman 2004). ESD therefore promotes a holistic perspective on sustainability that integrates ecological, economic, and social aspects. This means that unlike environmental education, sustainability education cannot be discussed within the context of one discipline or school subject alone; instead, it is usually said to require 'inter- or trans-disciplinarity' as in the UNESCO definition. Consequently, it is often described in a rather generic way that makes no reference to specific school subjects.

The same applies to sustainability problems, which are often described in a generic way because they are complex and multi-faceted real-world problems that require knowledge and concepts from many different domains to be properly understood, making it impossible to fully address them within the context of any individual school subject. In addition to their complexity, sustainability issues are often wicked, meaning that they are difficult or impossible to solve because any solution must satisfy a set of incomplete, contradictory, or changing requirements that are often difficult to distinguish. The concept of wicked problems was introduced by Rittel and Webber (1973) to describe issues that cannot be easily fixed, do not have a clear single solution, and show a certain resistance to resolution. The authors introduced the term to distinguish such problems from those usually addressed in disciplinary contexts, which they classify as 'tame' problems, that is, problems with a single, absolute solution. Sustainability problems are often wicked because they are symptoms of other problems that are generally strongly engrained into human society. Moreover, they are often inherently political problems because of the conflicting desires to improve the economy while protecting natural resources such as ecosystems. This means that attempts to address sustainability problems must often involve value-based decision-making about what aims to prioritise. To conclude, transformative sustainability education aims to develop citizens who have agency over sustainability issues as well as an understanding of the moral and political dimensions of sustainability problems (Öhman 2004; Sass et al. 2020).

The Turn Towards Competences in Education for Sustainable Development

The UNESCO definition of ESD only vaguely exemplifies content knowledge; instead, it describes the purpose and teaching approach of sustainability education. The policy text in which this definition was presented talks about 'key competencies' that should be developed by using 'learner-centred' teaching approaches of 'self-directed learning, participation and collaboration'. These approaches focus on the students (or learners) and developing their identity and their motivation to become agents of change in a sustainable way. ESD or sustainability education can thus be described as a way to develop learners' competences to handle complex and wicked problems.

The research literature on sustainability education similarly calls for development of students' competences (Vare et al. 2022) in areas including *systems thinking* (e.g., thinking about multidimensionality, addressing problems on multiple scales ranging from local to global, and feedback loops), *future thinking* (e.g., understanding deep time and scenario thinking), *values thinking* (e.g., risk analysis and ethical issues), *strategic competence* (e.g., organisational change and transformations) and *interpersonal competences* such as reflexivity, participation and teamwork (Wiek et al. 2011). The discourse on sustainability teaching and learning thus focuses less on imparting knowledge than on how the students as citizens can use their knowledge by taking action. The capacity to take action to address sustainability problems has been discussed in terms of competences or competencies (e.g., Wiek et al. 2011), skills (e.g., Rieckmann 2018), and capabilities (e.g., Nussbaum 2011). This shift towards competence-based education focusing on what is learned and what competences can be developed through schooling (i.e., the outcomes of learning) rather than knowledge-based education focusing on what is taught (the content) was named 'the competence turn' by Vare (2022) and is reflected in both policy and research on ESD. The main goal of the turn is to empower students to take and enact decisions about complex questions that lack easy solutions and where both values and facts must be considered.

In addition to the previously mentioned competence framework of Wiek et al. (2011), many ESD frameworks have been developed by policy makers in the European context including the Sustainable Development Model by the Environment and School Initiatives Organization (Sleurs 2008), and the OECD's Definition and Selection of Competences (Rychen and Salganik 2002). The more recent European sustainability competence framework GreenComp (European Commission 2022) defines four areas containing 12 competencies. Perhaps the most influential framework is the ESD Competences for Educators developed by the United Nations Economy Commission for Europe (UNECE 2012), which outlines 39 competences grouped into three broad categories: a holistic approach, envisioning change, and achieving transformation.

The above discussion can be distilled down to three conclusions. First, the generic competence-based education strategy that is often promoted in the literature aims to transform the learner into an action-oriented citizen. Second, the general aim of sustainability education is often described as fostering the development of action-competent citizens (Sass et al. 2020). Third, the concept of action competence has been defined to promote three main categories of interconnected sub-competences or dispositions in the learner: knowledge of action possibilities, confidence in one's own influence, and the willingness to act (Breiting and Mogensen 1999; Olsson et al. 2020).

The first of these sub-competences, knowledge of action possibilities, relates to knowledge about future actions (Bandura 2001), which requires skills in seeking information and building knowledge. The second sub-competence, confidence in one's own influence, relates to the empowerment of people and their self-confidence, enhancing their feeling of confidence and competence to act on the knowledge they possess. As such, this sub-competence requires self-efficacy and a feeling of being capable of dealing effectively with complex sustainability issues (Bandura 2001). The third sub-competence relates to the need for meaningful actions, which stems from people's desire and willingness to engage and make a difference (Kaplan and Kaplan 2009). Within the action competence concept, this is defined as a latent capacity among individuals; the need for meaningful actions is linked to the willingness to act for sustainability. If any one of these sub-components is missing, people will probably not act (Olsson et al. 2020). An action-competent person can thus be described as:

> [...] someone who is committed and passionate about solving a societal issue, has the knowledge about the issue at stake as well as about the democratic processes involved, takes a critical but positive stance toward different ways for solving it, and has confidence in their own skills and capacities for changing the conditions for the better. (Sass et al. 2020, p. 303)

On the basis of these considerations, transformative sustainability education has been put forward as a tool for developing students' action competence (Mogensen and Schnack 2010). Within this framework, promoting action competence is treated as an overarching goal of sustainability education that all teachers should strive for, regardless of subject specialty (UNESCO 2017). Ways of developing this competence in students may include acting in informed ways, participating in debates and discussions, and making decisions about developmental and future-oriented issues at both a private everyday level and a societal level (Jensen and Schnack 1997). Several authors recommend that such an approach should be coupled with authentic learning outside school and the confines of classroom-based subject-specific teaching (Sarid et al. 2024). This makes it difficult to situate the mission of developing students' action competence for sustainability within one school subject only and raises the question: which teachers are best placed to teach in a transformative way for action competence? In this chapter, it is argued that subject-specific competences from many disciplines are needed, and that language education teachers could make a unique and important contribution to this task.

Educational Purposes in Education for Sustainable Development

The renowned educational philosopher Gerd Biesta (2010, 2017) is one of the leading figures to have criticised 'the competence turn'. He argues that the turn towards generic skill-based curricula has caused a shift towards what he denotes as *learnification* of education, which he claims has undermined teaching and the role of the teacher (e.g., Biesta 2009, 2010, 2017). He further argues that this shift places the emphasis of education predominantly on the individual learner and their acquisition of skills and knowledge, in accordance with the preceding discussion of ESD. Biesta (2010, 2017) also argues that this competence shift is often driven by policies and practices that prioritise measurable outcomes such as standardised test scores over the intrinsic purposes of education (Biesta 2009). From this perspective, the call for action in ESD might imply some kind of evaluation and effects of education that have not traditionally been advocated for in educational policy or research. The call for action and 'the competence turn' thus create tensions because ESD changes the purpose of education.

Biesta identifies three primary functions of education that go beyond the content goals of specific school subjects and should be balanced to provide a holistic education (Biesta 2010, 2017). The first is *qualification*, which means equipping students with the knowledge, skills, and competencies needed to participate in society and working life. This function is performed within both academic and vocational education insofar as they seek to prepare students for future employment and societal roles. The second function is *socialisation*, which means integrating individuals into the social, cultural, and political fabric of society. This function helps students understand and adopt their community's norms, values, and traditions, fostering a sense of belonging and social cohesion. The third function is *subjectification*, which means guiding students to become independent, autonomous individuals. This involves encouraging students to think critically, question existing norms, and develop their own identities and perspectives (Biesta 2010, 2017).

Biesta's main critique is that learnification tends to de-emphasise the socialisation and subjectification functions of education. When education is reduced to a series of learning outcomes or competencies, there is a risk of neglecting students' development as whole persons who can engage critically with the world around them (Biesta 2010, 2017). Moreover, in a learnification-focused system, teachers may be seen primarily as facilitators of learning rather than as teachers who lead students through complex moral, social, and intellectual development processes.

One could argue that the critiques of 'the competence turn' and the focus on learning put forward by Biesta and others are actually against the caricaturisation or misuse of the competence concept by dividing it into too many individual skills or sub-competences, as exemplified by the 39 sub-competences of ESD defined by UNECE (UNECE 2012). Indeed, most of these frameworks directly acknowledge the problems of slicing educational goals into small pieces and seek to address these problems by specifying overarching levels of competence to encourage a more holistic perspective. For example, the UNECE ESD competences are aggregated into three main groups: holistic approach, envisioning change, and achieving transformation. This is consistent with the previously discussed complexity and wickedness of sustainability problems, which necessitate holistic and pluralistic perspectives (UNESCO 2017) as also argued by Biesta (2010, 2012) regarding education in general.

Perhaps the most obvious expression of the need for such perspectives in sustainability education and sustainability education research is the emphasis on generic action competence for sustainability, which encompasses multiple affective and cognitive dimensions that should be developed by the learner (Sass et al. 2020). The action competence concept emphasises the independent will to act, which Biesta would refer to as subjectification, but it also emphasises individual and collective functions. This is important because the collective functions relate directly to the socialisation function of education, meaning that the action competence concept promoted in transformative sustainability education relates directly to the functions of subjectification and socialisation. This contradicts Biesta's critique (2010, 2017) that these functions were neglected in the competence turn. Hence, by reorientating sustainability education towards the concept of action competence, rather than an innumerable number of sub-competences, a more holistic educational approach could be achieved that would not lead to learnification.

From this perspective, Biesta's critique of the competence turn is irrelevant to the arguments for more transformative competence-based education. However, it is not clear exactly which teachers should be expected to promote the desired competences in learners. Indeed, it is not immediately clear whether it is even possible for one teacher to address all the content dimensions, complex problems, and multiple competencies discussed above. Biesta (2017) asks some important questions about this issue, namely: what is the teacher's role in developing competences among learners and what professional competences do teachers need to be able to do this effectively? These questions are rarely addressed in the literature; instead, it is taken for granted that all teachers can enact ESD and implement the transformative perspective in their teaching and have a responsibility to do so.

When the issue of the teacher's role is addressed, interdisciplinary teaching approaches are often suggested in ESD (e.g., Munkebye et al. 2020). In this context, interdisciplinarity means incorporating perspectives from different disciplines or subjects into teaching about complex issues. However, it is not clear how this should be done. In a cross-curricular teaching approach, collaboration between teachers of different subjects can provide a complementary perspective on sustainability that leverages the individual strengths and disciplinary competences of teachers specialising in different subjects (Hudson 1995). The issue of how to enact interdisciplinary teaching in ESD will be addressed in Section 'The contribution of language education to ESD', but first, the issue of knowledge in education will be addressed.

Powerful Knowledge in Relation to Education for Sustainable Development

Like Biesta, the renowned educational philosopher Michael Young criticised 'the competence turn', albeit from a different perspective. Young (2008) coined the term *bring knowledge back in* when he drew on developments in the sociology of knowledge to critique the reorientation of formal schooling from learning disciplinary knowledge to developing skills or competences that students can use in their daily lives. Specifically, Young argued against education that lacks strong connections to formal academic knowledge. This criticism differs from Biesta's concerns about limiting the purpose of education to qualification and neglecting its other functions of socialisation and subjectification.

In his critique of 'the competence turn', Young introduced the concept of *powerful knowledge* (Young 2008), which relates to aspects of conceptual knowledge towards which teaching should be oriented. According to Young, powerful knowledge has two key characteristics. First, it is specialised in terms of how it is produced and transmitted in education, and this specialisation can be characterised in terms of the boundaries between disciplines and subjects. Young (2013) further stresses that powerful knowledge is not general knowledge and that these boundaries are not fixed and unchangeable. He also notes that cross-disciplinary research and learning depend on discipline-based knowledge. Second, powerful knowledge is differentiated from the experiences pupils bring to school by conceptual boundaries between school and everyday knowledge (Young 2013). True powerful knowledge is thus world-expanding and should empower students to make decisions and orient themselves in their own life-world.

Disciplinary knowledge, developed by 'communities of enquirers' (Young and Muller 2010, p. 14), is considered powerful because it is specialised, context-independent, and systematically principled (Young and Muller 2010). It achieves objectivity by being established with rigorous methodologies for generating and validating knowledge claims, and by being founded in concepts that are systematically interconnected within groups (Young 2015). Disciplinary knowledge is also considered powerful because it confers power to those who possess it (Gericke et al. 2018, 2022a), providing reliable explanations for observable phenomena, introducing new perspectives on the world, and offering a language for engaging in political, moral, and other debates (Young 2008). Acquiring disciplinary knowledge enables individuals to transcend their personal experiences and envision alternative possibilities (Muller and Young 2019).

Similarly to Biesta, Young emphasises the need for school education to address purposes beyond qualification alone, arguing for the importance of socialisation and especially subjectification, that is empowering students to become independent agents of change in society. However, Biesta (2010, 2017) focuses on problems resulting from slicing teaching up into uncontextualised pieces whereas Young's criticism focuses on a failure to address specialised academic subject knowledge.

Muller and Young identified two major traditions or trends in recent debates about curriculum and education that they consider deficient and problematic, and extrapolated two possible future trajectories arising from these traditions. In the first trajectory, denoted 'Future 1', knowledge is underpinned by an under-socialised epistemology (Young and Muller 2010) based on fixed sets of verifiable propositions or concepts that are evaluated in education, typically, through standardised testing. The problems identified with this approach align closely with the critique of Biesta (2010, 2017); Young and Muller (2010) argue that education in this tradition will not motivate students to engage with the body of knowledge to be learned and will thus fail to give students the knowledge needed to tackle the complex problems of contemporary society (Young 2015). The second trajectory, denoted 'Future 2', can be seen as a response to the first. Young and Muller (2010) claim that the epistemology of knowledge in 'Future 2' is *over-socialised* in that the character of knowledge is reduced to who knows and the identification of knowers and their practices and disciplinary knowledge is neglected. This critique also echoes that of Biesta (2010, 2017), who argues that the international turn towards generic skills-based curricula has resulted in the learnification of education, as previously discussed.

Because of these problems with the two main current educational traditions, Young and Muller (2010) proposed an alternative trajectory that they denote 'Future 3'. In their view, the most important part of curriculum making and teaching is to identify what constitutes powerful knowledge in different school subjects and to identify how such knowledge can be built within the boundaries of disciplines and school subjects. 'Future 3' features a subject-specific curriculum focused on giving students access to disciplinary knowledge that is fallible and open to change, shaped by the rules of specialist communities, and capable of empowering students to act as agents in their own life-world (Young 2013).

From an ESD or sustainability teaching perspective, the purpose of education is to be transformative, that is, to empower students to be change agents (UNESCO 2017) or equivalently, to develop their action competence for sustainability (Sass et al. 2020). Based on Young's description (2013) of powerful knowledge and teaching in the 'Future 3' trajectory, this purpose aligns closely with the subjectification function outlined by Biesta (2010, 2017). Drawing on Young's (2008, 2015) ideas about powerful knowledge, one could thus argue that subject-specific education could be an important avenue for subjectification-oriented ESD aiming to develop students' action competence for sustainability. This raises a question: how can transformative ESD teaching with a subject-specific orientation be implemented and what role could language teaching play in such a process?

The Contribution of Language Education to Education for Sustainable Development

As shown in the preceding sections, sustainability education has been described as a progressive and transformative teaching approach that focuses on the learner and seeks to develop learners' competencies for sustainable action. This is prioritised in many national curricula, such as Finland (EDUFI 2014), Norway (NME 2017), and Sweden (SNAE 2018), where sustainability is presented as an overriding goal that all teachers should seek to address no matter what subject they teach.

However, this idealistic way of presenting sustainability education as a responsibility of all teachers is not consistent with the ways in which teachers actually experience its relevance: sustainability has been integrated effectively into the teaching traditions of some school subjects but not in others (Stables and Scott 2002). For example, science and geography teachers recognise sustainability education as an extension of environmental education, which is part of the core curriculum of their subjects (Breiting 2000), whereas language teachers have tended to avoid teaching sustainability: studies conducted in Nordic countries have shown that language teachers address sustainability issues in their teaching less commonly than teachers of other subjects (Borg et al. 2012; Sund and Gericke 2020; Uitto and Saloranta 2017). One reason for this is that language teachers tended not to see such issues as being relevant to language teaching (Borg et al. 2012), and instead saw their role as complementing the sustainability education provided within other subjects by giving students communication and problem-solving skills as well as procedural knowledge that are relevant in sustainability education (Sund and Gericke 2020; Uitto and Saloranta 2017). Language teachers dealt mostly with sociocultural issues (e.g., cultural heritage and identity, and multiculturalism) and seldom considered ecological and economic issues (Borg et al. 2014; Uitto and Saloranta 2017). The few existing empirical studies on sustainability education in the context of language education thus indicate that language teachers consider their role in sustainability education to be limited and focused mainly on supporting teachers of science and social science. In opposition to such a view, this volume seeks to demonstrate the vital role of language education in enacting and promoting transformative sustainability education.

The chapters in this volume will argue for the importance of language education in ESD and show how language education can support and be part of transformative sustainability education. This chapter has shown that ESD and sustainability education have mainly been characterised as general teaching approaches in the policy and research literature, with limited efforts to address disciplinary or subject-specific aspects of teaching. Biesta (2010, 2017) and Young (2008, 2013) have problematised this way of perceiving education, showing that it can promote learnification and give rise to either over- or under-socialised epistemology. However, both groups of authors acknowledge the importance of the socialisation and subjectification functions of education, which are key areas of focus in transformative sustainability education that seeks to develop students' action competence for sustainability. This chapter argued that subject-specific education as envisioned in the 'Future 3' trajectory (Young and Muller 2010) has an important part to play in achieving this goal, with language education having a significant role. Next, I will summarise the findings of an empirical study that investigated how language education can contribute to transformative sustainability education alongside science and social science studies (Sund and Gericke 2020; Sund et al. 2020).

The study in question used a research design based on focus group interviews with 43 lower secondary school teachers specialising in different subjects, who were asked to discuss their contributions to ESD or sustainability education based on their backgrounds as teachers of natural science, social science, and language (including literature). The 43 participating teachers were divided into ten groups based on school and subject area, with three to ten teachers in each group. The focus group discussions were fully transcribed and analysed based on the research questions of what, how, and why the participants felt that their teaching could contribute to ESD (Sund and Gericke 2020; Sund et al. 2020). Analysis of the transcripts revealed aspects of ESD teaching common to these different subject areas as well as some contributions unique to each subject. The study's findings were presented in detail by Sund and Gericke (2020) and can be summarised as follows.

The science teachers primarily considered their main contribution to sustainability education (as well as their main goal in this context) to be conveying scientific facts, particularly relating to environmental issues, such as the consequences of energy use and generation, eutrophication, pollution, and global warming. Moreover, they promoted a teacher-centred lecturing approach whose main purpose is to transmit scientific knowledge and skills in scientific inquiry (Sund and Gericke 2020). Social science teachers felt that their purpose in sustainability education was to enable action competence in societal issues. Their perceived contribution primarily related to their disciplinary traditions concerning social justice, international relations, politics, trade, and peace. The social science teachers' approach to teaching differed from the science teachers' methods, often incorporating critical investigation and group discussions about the consequences of climate change and social justice in human relations at both individual and collective levels (Sund and Gericke 2020).

The language teachers considered their main contribution to ESD to be their long-term influence on students' personal development and identity formation, enabling students to feel capable, empowered, and encouraged to participate as independent agents with the ability to influence societal development. They felt that this could be achieved by having students read authentic literature that could foster their inner development and identity, and by communicating about sustainability issues. The language and literature teachers did not think that there was any content specific to their own discipline that was important in relation to sustainability education; instead, they often used media material such as books, news, and TV channels as sources of content in order to make their teaching close to the students' everyday lives as experienced through media. The language teachers often relied on information and communication technology tools for self-study and group work, which meant their lessons could be connected to the world outside school. In addition, they aimed to help students practice and develop their communication competences (Sund and Gericke 2020).

The science and social science teachers' sustainability teaching had common features because topics such as climate change issues were included in both groups' curricula. These two teacher groups felt that they had the most in common, making it easy for them to collaborate within ESD using multidisciplinary perspectives. The language teachers felt they had less in common with the other teacher groups but that they could collaborate via media content and competence development practices such as oral communication (for more details, see Sund and Gericke 2020). This shows that there are overlaps that can serve as starting points for collaboration within multidisciplinary sustainability teaching.

Transformative Language Teaching for Sustainability

The most interesting findings of Sund and Gericke (2020) were that literature and language teachers highlighted the potential of their teaching to develop students' agency and guide their identity formation, allowing them to feel capable of participating in the general development of society and encouraging them to do so. In language and literature education, the disciplinary tradition is not to teach predefined concepts as in science and social science education. Instead, besides developing skills such as reading and argumentation, the core purpose of the subject is self-formation, helping students develop their self-identity and self-efficacy, and to become independent actors in society (Darvin and Norton 2023). This subject-specific purpose of literature and language education resonates well with the idea of action competence, which requires teachers to address affective issues in order to boost students' willingness to engage in sustainability issues, and to build their self-efficacy to convince them that their actions genuinely matter (Mogensen and Schack 2010; Sass et al. 2020).

Sund and Gericke (2020) showed that language and literature education have a subject tradition that can fulfil these important functions, and that language teachers have the professional competence needed to enact these functions. The inclusion of language teaching in ESD thus provides a way to implement the educational functions of socialisation and especially subjectification, which Biesta (2010, 2017) noted may be easily overlooked as a result of 'the competence turn'. Indeed, multidisciplinary teaching based on complementary subject-specific perspectives, including those originating from language teaching, could be an important tool for enacting transformative sustainability teaching. In such a multidisciplinary approach, powerful knowledge drawn from different school subjects and disciplines would be combined to empower students with action competence for sustainability. Hence, if we look beyond the content perspective in sustainability teaching and instead view teaching from a more holistic perspective, different subject traditions can contribute uniquely to ESD.

Returning to the 'Future 1 and 2' curriculum trajectories described by Young and Muller (2010), the former produces an under-socialised epistemology with no connection to authentic issues or students' life-worlds, both of which are central to sustainability education. Conversely, the latter leads to an over-socialised epistemology that does not provide the subject-specific knowledge needed to support the development of action competence. Instead, sustainability education could follow the third trajectory described by Young and Muller, 'Future 3', in which the differentiation of knowledge between different school subjects is recognised and exploited. This could be important for sustainability education because different disciplines have different knowledge structures and ways of teaching. Thus, exposure to teachers and teaching rooted in different disciplines could help instil the variation and open-endedness needed to achieve the purpose of ESD or sustainability education.

Language education confers powerful knowledge that serves the functions of both qualification and socialisation, but perhaps even more importantly, it also contributes to the function of subjectification by developing students' action competence. This is important because teachers of natural and social science focus mainly on teaching concepts and factual knowledge in their sustainability teaching, meaning that their ESD work contributes relatively little to subjectification. Conversely, the ESD efforts of language teachers seem to focus more on developing learners' action competence for sustainability, meaning that they could contribute significantly to subjectification (Sund and Gericke 2020).

The above argument indicates that the main contribution of transformative language education to sustainability education is in the *power* dimension of *powerful knowledge* rather than the knowledge dimension. White (2019) has previously discussed the problem of defining core concepts in disciplines and school subjects outside of science and mathematics in relation to powerful knowledge. However, in light of 'the competence turn', the main contribution of transformative language education to sustainability education might actually be the empowerment it can confer. As shown in a review of the expression of powerful knowledge in several school subjects (Hudson et al. 2022), language education focuses on the learner rather than the academic discipline (Gericke et al. 2022b). Consequently, powerful knowledge in language education is closely linked to a core objective of transformative education: to encourage learners to reflect on taken-for-granted ways of viewing the world and to encourage fundamental shifts in how they see and understand themselves and their relationship with the world (Mezirow 2000).

These ideas were developed further in a conceptual article that investigated the possibilities offered by language education in the context of sustainability education and proposed a transformation-oriented ESD framework that could be foundational in transformative language teaching for sustainability (Maijala et al. 2024). The framework outlines linguistic and cultural features of sustainability that are central to language teaching and learning. It also addresses pedagogical issues—in particular, subject matter content and methods for transformative language teaching in sustainability education that enable language teachers to integrate ESD into their lessons using learner-oriented language teaching methods (Maijala et al. 2024). The rest of this volume extends these ideas, and the potentially important contribution of transformative language education to sustainability education and how it can be developed from theory to practice.

References

Bandura, A. 2001. Social cognitive theory: An agentic perspective. *Annual Review of Psychology* 52 (1) (February): 1–26. https://doi.org/10.1146/annurev.psych.52.1.1.

Biesta, G. 2009. Good education in an age of measurement: On the need to reconnect with the question of purpose in education. *Educational Assessment, Evaluation and Accountability* 21 (1) (February): 33–46. https://doi.org/10.1007/s11092-008-9064-9.

Biesta, G. 2010. *Good education in an age of measurement: Ethics, politics, democracy*. London: Paradigm Publishers.

Biesta, G. 2017. *The rediscovery of teaching*. London and New York: Routledge. https://doi.org/10.4324/9781315617497.

Borg, C., N. Gericke, H.-O. Höglund, and E. Bergman. 2012. The barriers encountered by teachers implementing Education for Sustainable Development: Discipline bound differences and teaching traditions. *Research in Science & Technological Education* 30 (2) (July): 185–207. https://doi.org/10.1080/02635143.2012.699891.

Borg, C., N. Gericke, H.-O. Höglund, and E. Bergman. 2014. Subject- and experience-bound differences in teachers' conceptual understanding of sustainable development. *Environmental Education Research* 20 (4) (July 4): 526–551. https://doi.org/10.1080/13504622.2013.833584.

Breiting, S. 2000. Sustainable development, environmental education and action competence. In *Critical Environmental and Health Education*, vol. 46, ed. B.B. Jensen, K. Schnack, and V. Simovska, 151–166. Copenhagen: Research Centre for Environmental and Health Education. The Danish University of Education.

Breiting, S., and F. Mogensen. 1999. Action competence and environmental education. *Cambridge Journal of Education* 29 (3) (November): 349–353. https://doi.org/10.1080/0305764990290305.

Darvin, R., and B. Norton. 2023. Investment and motivation in language learning: What's the difference? *Language Teaching* 56 (1) (January): 29–40. https://doi.org/10.1017/S0261444821000057.

Eilam, E., and T. Trop. 2010. ESD pedagogy: A guide for the perplexed. *The Journal of Environmental Education* 42 (1) (September 17): 43–64. https://doi.org/10.1080/00958961003674665

EDUFI (Finnish National Agency for Education). 2014. National core curriculum for basic education 2014, 3rd Edition. Opetushallitus, 2020.

European Commission. Joint Research Centre. 2022. GreenComp, the European sustainability competence framework. LU: Publications Office. https://doi.org/10.2760/13286.

Gericke, N. 2022. Teaching for the Anthropocene: Bildung-oriented Education for Sustainable Development in a subject-specific curriculum. In *Bildung, knowledge, and global challenges in education*, ed. E. Krogh, A. Qvortrup, S. T. Graf, 53–69. London and New York, Taylor & Francis. https://doi.org/10.4324/9781003279365-5.

Gericke, N., B.J. Hudson, C. Olin-Scheller, and M. Stolare. 2018. Powerful knowledge, transformations and the need for empirical studies across school subjects. *London Review of Education* (November 15). Available from: https://scienceopen.com/document?vid=b7ff983e-a0b1-4c3e-a3bb-2f183505257c (accessed October 22, 2024).

Gericke, N., A. Manni, and U. Stagell. 2020. The green school movement in Sweden—Past, present and future. In *Green schools movements around the world: Stories of impact on Education for Sustainable Development*, ed. A. Gough, J. Chi-Kin Lee, E. P. K. Tsang, 309–332. Cham, Switzerland: Springer. https://doi.org/10.1007/978-3-030-46820-0_17.

Gericke, N., H. Brian, C. Olin-Scheller, and M. Stolare. 2022a. Researching powerful knowledge and epistemic quality across school subjects. In *International perspectives on knowledge and curriculum: Epistemic quality across school subjects*, ed. B. Hudson, N. Gericke, C. Olin-Scheller, and M. Stolare, 1–16. London: Bloomsbury Publishing.

Gericke, N., H. Brian, C. Olin-Scheller, and M. Stolare. 2022b. Trajectories of epistemic quality and powerful knowledge across school subjects. In *International perspectives on knowledge and curriculum: Epistemic quality across school subjects*, ed. B. Hudson, N. Gericke, C. Olin-Scheller, and M. Stolare, 197–221. London: Bloomsbury Publishing.

Hudson, B. 1995. Group work with multimedia in the secondary mathematics classroom (Doctoral thesis), Hallam University, Sheffield, United Kingdom.

Hudson, B., N. Gericke, C. Olin-Scheller, and M. Stolare, ed. 2022. *International perspectives on knowledge and quality: Implications for innovation in teacher education policy and practice*. Reinventing Teacher Education/Series, ed. M. Brennan, V. Ellis, J. Nuttall, and P. Smagorinsky, 1st ed. New York; London: Bloomsbury Academic.

Jensen, B.B., and K. Schnack. 1997. The action competence approach in environmental education. *Environmental Education Research* 3 (2) (May): 163–178. https://doi.org/10.1080/1350462060094305 3.

Kaplan, S., and R. Kaplan. 2009. Creating a larger role for environmental psychology: The reasonable person model as an integrative framework. *Journal of Environmental Psychology* 29 (3) (September): 329–339. https://doi.org/10.1016/j.jenvp.2008.10.005.

Kollmuss, A., and J. Agyeman. 2002. Mind the gap: Why do people act environmentally and what are the barriers to pro-environmental behavior? *Environmental Education Research* 8 (3) (August): 239–260. https://doi.org/10.1080/13504620220145401.

Maijala, M., N. Gericke, S.-R. Kuusalu, L.M. Heikkola, M. Mutta, K. Mäntylä, and J. Rose. 2024. Conceptualising transformative language teaching for sustainability and why it is needed. *Environmental Education Research* (February 19): 1–20. https://doi.org/10.1080/13504622.2023.2167941.

Marcinkowski, T., and A. Reid. 2019. Reviews of research on the attitude-behavior relationship and their implications for future environmental education research. *Environmental Education Research* 25 (4) (April 3): 459–471. https://doi.org/10.1080/13504622.2019.1634237.

Mezirow, J., ed. 2000. *Learning as transformation: Critical perspectives on a theory in progress.* The Jossey-Bass higher and adult education series. San Francisco: Jossey-Bass.

Mogensen, F., and K. Schnack. 2010. The action competence approach and the "new" discourses of Education for Sustainable Development, competence and quality criteria. *Environmental Education Research* 16 (1) (February): 59–74. https://doi.org/10.1080/13504620903504032.

Muller, J., and M. Young. 2019. Knowledge, power and powerful knowledge re-visited. *The Curriculum Journal* 30 (2) (June): 196–214. https://doi.org/10.1080/09585176.2019.1570292.

Munkebye, E., E. Scheie, A. Gabrielsen, A. Jordet, S. Misund, T. Nergård, and A.B. Øyehaug. 2020. Interdisciplinary primary school curriculum units for sustainable development. *Environmental Education Research* 26 (6) (June 2): 795–811. https://doi.org/10.1080/13504622.2020.1750568.

NME. Norwegian Ministry of Education. 2017. Overordnet del—verdier og prinsipper for grunnopplæringen [Overarching part—Values and principles of primary and secondary education]. Utdanningsdirektoratet.

Nussbaum, M.C. 2011. *Creating capabilities: The human development approach.* Cambridge: Harvard University Press.

Öhman, J. 2004. Moral perspective in selective traditions of environmental education. In *Learning to change our world*, ed. P. Wickenberg, H. Axelsson, L. Fritzén, G. Helldén, J. Öhman, 33–57. Lund: Studentlitteratur.

Olsson, D., N. Gericke, W. Sass, and J. Boeve-de Pauw. 2020. Self-perceived action competence for sustainability: The theoretical grounding and empirical validation of a novel research instrument. *Environmental Education Research* 26 (5) (May 3): 742–760. https://doi.org/10.1080/13504622.2020.1736991.

Rieckmann, M. 2018. Learning to transform the world—Key competencies in Education for Sustainable Development. *Issues and Trends in Education for Sustainable Development* 39 (1): 39–59.

Rittel, H.W., and M.M. Webber. 1973. Dilemmas in a general theory of planning. *Policy Sciences* 4 (2): 155–169. https://doi.org/10.1007/BF01405730

Rychen, D.S., and L.H. Salganik. 2002. Definition and selection of competencies (DeSeCo): Theoretical and conceptual foundations. Neuchatel, Switzerland: Swiss Federal Statistical Office.

Sarid, A., J. Boeve-de Pauw, A. Christodoulou, M. Doms, N. Gericke, D. Goldman, P. Reis, A. Veldkamp, S. Walan, and M.C.P.J. Knippels. 2024. Reconceptualizing open schooling: Towards a multidimensional model of school openness. *Journal of Curriculum Studies* (August 23): 1–19. https://doi.org/10.1080/00220272.2024.2392592.

Sass, W., J. Boeve-de Pauw, D. Olsson, N. Gericke, S. De Maeyer, and P. Van Petegem. 2020. Redefining action competence: The case of sustainable development. *The Journal of Environmental Education* 51 (4) (July 3): 292–305. https://doi.org/10.1080/00958964.2020.1765132.

Sleurs, W. 2008. Competencies for ESD (Education for Sustainable Development) teachers. A framework to integrate ESD in the curriculum of teacher training institutes. Available from: https://unece.org/fileadmin/DAM/env/esd/inf.meeting.docs/EGonInd/8mtg/CSCT%20Handbook_Extract.pdf (accessed October 2024).

Smyth, J.C. 2006. Environment and education: A view of a changing scene: Reprinted from *Environmental Education on Research* (2005) 1 (1): 3–20. *Environmental Education Research* 12 (3–4) (July): 247–264. https://doi.org/10.1080/13504620600942642.

Stables, A., and W. Scott. 2002. The quest for holism in Education for Sustainable Development. *Environmental Education Research* 8 (1) (February): 53–60. https://doi.org/10.1080/13504620120109655.

Sund, P., and N. Gericke. 2020. Teaching contributions from secondary school subject areas to Education for Sustainable Development—A comparative study of science, social science and language teachers. *Environmental Education Research* 26 (6) (June 2): 772–794. https://doi.org/10.1080/13504622.2020.1754341.

Sund, P., N. Gericke, and G. Bladh. 2020. Educational content in cross-curricular ESE teaching and a model to discern teacher's teaching traditions. *Journal of Education for Sustainable Development* 14 (1) (March): 78–97. https://doi.org/10.1177/0973408220930706.

SNAE. Swedish National Agency for Education. 2018. Curriculum for the compulsory school, preschool class and school-age educare 2011. Skolverket.

Uitto, A., and S. Saloranta. 2017. Subject teachers as educators for sustainability: A survey study. *Education Sciences* 7 (1) (January 4): 8. https://doi.org/10.3390/educsci7010008.

UNESCO. 2012. Learning for the future: Competences in Education for Sustainable Development. Geneva: United Nations. Available from: https://unece.org/fileadmin/DAM/env/esd/ESD_Publications/Competences_Publication.pdf (accessed October 22, 2024).

UNESCO. 2017. *Global Action Programme on Education for Sustainable Development*. Paris: UNESCO.

Vare, P. 2022. The competence turn. In *Competences in Education for Sustainable Development: Critical perspectives*, ed. P. Vare, N. Lausselet, and M. Rieckmann, 11–18. Cham: Springer International Publishing.

Vare, P., N. Lausselet, and M. Rieckmann. 2022. Introduction. In *Competences in Education for Sustainable Development: Critical perspectives*, ed. P. Vare, N. Lausselet, and M. Rieckmann, 3–10. Cham: Springer International Publishing.

White, J. 2019. The end of powerful knowledge? *London Review of Education* 17 (3). Available from: https://doi.org/10.18546/LRE.17.3.15 (accessed October 22, 2024).

Wiek, A., L. Withycombe, and C.L. Redman. 2011. Key competencies in sustainability: A reference framework for academic program development. *Sustainability Science* 6 (2) (July): 203–218. https://doi.org/10.1007/s11625-011-0132-6.

Young, M. 2008. *Bringing knowledge back in: From social constructivism to social realism in the sociology of education.* UK, Routledge.

Young, M. 2013. Overcoming the crisis in curriculum theory: A knowledge-based approach. *Journal of Curriculum Studies* 45 (2) (April): 101–118. https://doi.org/10.1080/00220272.2013.764505.

Young, M. 2015. Powerful knowledge as curriculum principle. In *Knowledge and the future school: Curriculum and social justice*, ed. M. Young, D. Lambert, C. Roberts, and M. Roberts, 65–88, 2nd ed. London, Bloomsbury Publishing.

Young, M., and J. Muller. 2010. Three educational scenarios for the future: Lessons from the sociology of knowledge. *European Journal of Education* 45 (1) (March): 11–27. https://doi.org/10.1111/j.1465-3435.2009.01413.x.

3

Linking Education for Sustainable Development Globally with Language Learning

Salla-Riikka Kuusalu

Introduction

Throughout the past few decades, aspects of sustainability have been incorporated into the curricula of various language-related subjects, such as foreign language studies, English literature, translation studies, and linguistics (see Canning 2010), but only recently have systematic approaches on how to teach about sustainability in language classes emerged (see, e.g., Melin 2019; de la Fuente 2021; Maijala et al. 2024). Our understanding of languages, especially the ones used when discussing about the environment, sustainability, or sustainable development, directly influences our literacy in the given subjects (Canning 2010). Thus, Education for Sustainable Development (ESD) cannot have a solid foundation unless learners have sufficient literacy in the

S.-R. Kuusalu (✉)
School of Languages and Translation Studies, University of Turku, Turku, Finland
e-mail: srkbos@utu.fi

Department of Biology, University of Turku, Turku, Finland

M. Maijala et al. (eds.), *Transformative Language Teaching for Sustainability*,
https://doi.org/10.1007/978-3-031-85493-4_3

main concepts of sustainable development. Transformative approaches addressing the issue, such as ecopedagogy (Misiaszek 2020), aim at building up this literacy through, e.g., dialogue and critical (re)thinking. Another aspect worth noticing is the transformative potential of Indigenous narratives and social representations and their inherent conceptualisations of concepts related to sustainable development, many of which differ profoundly from the Global North perspective (Virtanen et al. 2020).

From the learner's viewpoint, learning of complex concepts is most effective through their first or most used home language(s) (e.g., Litre et al. 2022). Therefore, teaching about sustainable development would often require the use of alternative teaching methods, such as translanguaging (e.g., Creese and Blackledge 2010) or other dynamic multilingual approaches (see, e.g., Creese and Martin 2003; García and Wei 2015; Makalela 2017) in all ESD including in language lessons. This becomes essential in the large and variable context of the Global South, where multilingualism is prevalent and the ongoing trend of using global languages in communicating about sustainable development may even hinder critical sustainability transitions (Litre et al. 2022). In the African continent, the development of learners' linguistic capabilities is challenged by dominantly Eurocentric curricula and teachers' insufficient multilingual capabilities and pedagogical skills in developing multilingual capabilities in learners (Tikly 2019; Norro 2021). In addition, as language itself is a complex system (Beckner et al. 2009), it is beneficial for both teachers and learners to understand how linguistic processes function and how they affect learning languages and learning complex concepts in foreign language lessons.

Lupele and Lotz-Sisitka (2014) found in their report for UNESCO of ESD in sub-Saharan Africa that the effective learning processes used in ESD have the potential to increase the quality of education and make it more relevant to both teachers and learners. In European language education, ESD and the Common European Framework of Reference for Languages (CEFR; Council of Europe 2001) already share many teaching practices supporting the same competencies (e.g., student-centred activities, problem-solving exercises, and communicative tasks), which also support important ESD learning processes such as action competence and critical thinking. The gradual transition of education towards a more learner-centred approach with more emphasis on peer interaction and experiental learning (Agbedahin 2019) has influenced the development of transformative ESD approaches in general (e.g., Vare and Scott 2007; Sinakou et al. 2019) and those directed for language education (Maijala et al. 2024). In this chapter, the perspective of the learner is highlighted, and the focus is on two main questions: how ESD in language education affects language learning both in the Western world and in the context of the Global South and what is required from ESD in language lessons to enhance learning that supports transformative actions.

Learning Complex Adaptive Systems—Languages

Language learning can be considered unique in the way that the content of learning and its medium of instruction are often the same (Hammadou and Bernhardt 1987; Halliday 1993), which can pose challenges specifically to inexperienced or novice learners. Unlike the traditional, grammar-based, and school-oriented approach to language as a relatively stable and constant system, modern approaches to language view it as a highly fluid and adaptive system, which is influenced by the past, present, and future factors in individual language user's life, as well as the societies and environments languages are used in (Beckner et al. 2009). Based on research conducted in the areas of language change, use, acquisition, and related computer modelling, Beckner and others (2009) distinguished seven characteristics that language, viewed as a complex system, possesses:

1) Language emerges both as an idiolect via social interaction and as a communal language via the interaction of the idiolects. Thus, its control is distributed and its emergence collective.
2) There is no ideal language user, representation, or development type. This leads to inherent diversity and heterogeneity of the language and its use.

3) Language is an open and dynamic system experiencing perpetual changes.
4) Language is adaptive: it changes through the competition of linguistic elements and diverse feedback systems in social interaction.
5) Small quantitative differences in certain linguistic parameters may lead to rapid and even escalating phase transitions in the quality of information. As an example, words may change their meaning and grammatical positions over time.
6) Social networks limit and influence language use and linguistic interactions.
7) The complexity of language arises gradually from the cultural adaptation of the mind to social human interactions.

When considering language learning from the perspective of these characteristics, learners become active participants of the language learning process (Larsen-Freeman 2019), i.e., they have more agency on their learning. This changes the position of foreign/second language learners from passive and top-down controlled receivers of linguistic knowledge into capable, self-learning individuals, who are 'optimising conditions for one's own learning and choosing to deploy one's semiotic resources to position one-self as one would wish in a multilingual world' (Larsen-Freeman 2019, p. 62). This approach has been conceptualised in the Complex Dynamic Systems Theory (CDST) by Larsen-Freeman (2019), which suggests further, that to understand singular learning paths, one needs to explore the individual's interactions with their environment. Active participation of language users is also highlighted in the views of Kramsch (2014), who describes multilingual surroundings as places of constant choice-making, where meanings of utterances are negotiated between the people involved and where language is tailored for specific people and situations, which may be quite different from one's own frame of reference. Moreover, languages themselves can have connotations related to the history of their users, such as being the language of former oppressors, or on the other hand, being one's cherished heritage language, and these connotations affect which language is chosen by the language user for each situation in question (Kramsch 2014). From these perspectives, language learners are both free to be active in their learning process, but simultaneously affected by the historical weight bestowed upon the language itself, not to mention the expectations of other language users.

In the multidisciplinary field of second language acquisition (SLA), the importance of individual behaviours and aspirations of learners to the success of language learning has been argued for by Papi and Hiver (2024), who proposed a novel perspective to the social-cognitive second language (L2) learning theory that puts the language learner at focus in the learning process: the proactive language learning theory (PLLT). The theory includes the main SLA concepts (input, interaction, form-focused instruction, and corrective feedback) and proposes four key dimensions to proactive L2 learning: the behaviours of input-seeking, interaction-seeking, feedback-seeking, and information-seeking (Papi and Hiver 2024). According to the researchers, these four dimensions are interrelated, dynamic, can be simultaneous and synergetic, and are most likely to account for differences in language learning success. The main argument that differs from the prevailing SLA theories is that PLLT is proposing that 'learners deliberately and selectively employ these cognitive mechanisms to use the L2 opportunities that they seek, find and engage with' (Papi and Hiver 2024, p. 17). If proven valid by further research, this theory would support the notion that if ESD makes students personally interested in the tasks and themes used in language classes and especially if they show proactive behaviours because of it, it is likely that ESD could improve their success also in learning the target language.

In language education, two major paradigm shifts have recently occurred: First, communicative approaches (e.g., Jacobs and Farrell 2003) started to replace the earlier methods based on grammar and translation; Consecutively, transformative language learning and teaching approaches have begun to revolutionarise the field (Leaver et al. 2021). Somewhere in between these two movements, intercultural approaches such as the model developed by Liddicoat and Scarino (2013) emerged. In essence, the position of the language learner has shifted from being an actor/interactor in the process of learning towards what Leaver (2021) describes as the role of a client, whereas teachers could be viewed as counsellors who guide the way instead of determining a ready path for each learner. Further still, intercultural language learning (Liddicoat and Scarino 2013) encompasses the idea that interpretation is at the core of understanding languages and the process of learning is reciprocal in nature. What do these changes mean for the learning of second or foreign languages?

To start with, the goal of learning to use language in communication changes towards an idea of the learner developing into an increasingly competent user of the language and its multilingual repertoires (Léglise 2017). The information for learning the language can come from multiple sources, not just the teacher, but also from society, culture, media, and the learners themselves. These paradigm shifts also entail a change in the use of learners' cognitive functions as they involve being more aware of oneself, others, and one's environment, being able to think critically and innovatively and to reframe existing ways of thinking (Leaver 2021). Adding intercultural language learning into the equation gives prominence to the interpretations learners make of themselves, their fellow learners, and of their own learning (Liddicoat and Scarino 2013), allowing their voices to be heard and appreciated during the language learning process. The most eminent result these changes in language education are proposing lies in their empowering potential: Would language learning become easier and more pleasant for everybody, if learners would be given more power and importance in their own language learning?

Naturally, each paradigm shift is followed by the refinement of old and development of new theories, approaches, and practical applications, and their parallel emergence may cause diverging paths in both research and practice. As an example, Oxford (2021) has elaborated on the seemingly different approaches to transformative learning by Jack Mezirow and John M. Dirkx, the former promoting a cognitive-analytic and the latter an emotional-integrative approach. Through the concepts such as perspective transformation, meaning schemes, and reflection, Mezirow (1978) constructed his approach, which received critique, e.g., for undermining emotional, personal, social, and political factors that affect transformations and pointedly, from their view that collectivist cultures would not be able to change their perspectives (Baumgartner 2012, in Oxford 2021). According to Oxford (2021), Dirkx's (2001) approach has its roots in depth psychology, especially Jungian theory, and it involves, e.g., the concepts of *ego consciousness*, *personal* and *transpersonal unconscious,* and *individuation*: In stark contrast to Mezirow's rational approach, Dirkx suggested that emotions were at the heart of transformation. However, the conclusion of Oxford (2021) was not to choose one over the other, but to embrace the interlinkage of cognition and emotion and use the different approaches to enable transformative learning for a wider learner base.

In conclusion, languages viewed as complex adaptive systems and the paradigm shifts leading language learning towards transformative approaches align nicely with the transformative goals of ESD and its learning processes that can even enhance the personal and social relevance of language lessons (Lupele and Lotz-Sisitka 2014). Next, these learning processes will be examined more closely in the context of Sub-Saharan Africa, drawing a line towards multilingual realities growing in number around the world.

Learning Processes of Education for Sustainable Development

ESD has been developed to enable UNESCO's (2018) Agenda 2030 Action Plan and advance the Sustainable Development Goals (SDGs) through educators. The underlying idea is that education provides a social context for societal transformation, and if educators consider ESD to be of value to their teaching, they can support the empowerment of learners as part of this process (Lupele and Lotz-Sisitka 2014). ESD learning has also been associated with the ability of individuals and communities to respond and prepare for future risks (Mukute and Lotz-Sisitka 2012). Prior to the launch of Agenda 2030, UNESCO commissioned an expert review on the commonly accepted and commensurate learning processes that could be promoted via ESD activities and on the frameworks related to these (Tilbury 2011). The report concludes that the processes which form the basis of ESD are collaboration and intra- and intercultural dialogue, holistic engagement, innovative approaches, and experiental teaching and learning, as well as active and participatory learning. Hence, instead of just learning about sustainability or sustainable development, ESD engages with action competence: learning to think critically, ask relevant questions, envision sustainable futures, use applied learning to respond to challenging situations, and use dialogue in negotiating solutions between traditional knowledge and innovation (Tilbury 2011). These learning processes correspond for a large part with the transformative language learning processes already emerging in different fields of language education (Leaver 2021) and make language lessons a highly potential outlet for ESD.

In their research report to UNESCO, Lupele and Lotz-Sisitka (2014) reviewed the literature and several case studies of implemented ESD learning processes and analysed contextually the factors surrounding these processes and their conceptualisation in the African continent. The aim of the study was to analyse how the learning processes of ESD contribute to the quality and relevance of education. Lupele and Lotz-Sisitka (2014) concluded that ESD learning processes have several implications for the way educators think about the quality of education, such as shifting the focus of learning into more inclusive meaning making via inter-epistemological dialogue, increasing critical thinking, action competence, and agency in learners, making learning more relevant by engaging learning with locally meaningful initiatives and connecting them with the wider community and the world. Most importantly, quality of education was seen to improve through the learning processes of ESD, because they bring forth culturally and socially relevant aspects such as social justice and inclusiveness of education and give an alternative framing to educational quality (Lupele and Lotz-Sisitka 2014).

Thus, in addition to language teaching being able to promote ESD, ESD could contribute to the quality of language education by expanding it from the current language learning processes towards a more holistic and inclusive understanding of language and its use and increasing multiple sustainability competencies of language learners through meaningful activities both in the classroom and within the society.

Transferable Sustainability Competencies

The European Sustainability Competence Framework GreenComp (European Commission 2022) was developed to guide educators in their work in teaching about environmental sustainability and to help all learners to develop their skills and knowledge in how to live sustainably, i.e., respecting and prioritising all life forms and making sure that human activity does not exceed planetary boundaries. In the framework, a sustainability competence is described as a competence that empowers learners to realise and incorporate sustainability values and manage dealing with complex systems, so that they can act responsibly in the society and work towards a sustainable future for the planet. The framework further describes various competencies related to areas of sustainability values and complexity, sustainable futures, and sustainable actions. Such competencies are transferable in the sense that skills and knowledge gathered in one area of life can be used to help matters in other areas. On the other hand, when we consider language competencies, many of them transfer effortlessly into sustainability competencies, given the enormous power of language in shaping how we think and act (see, e.g., Stibbe 2015, 2024). Some of these competencies, such as action competence, systems thinking, and critical thinking, were already touched briefly when discussing learning processes that support ESD, but language learning also involves several transferrable competencies that are described in the CEFR (Council of Europe 2001).

After the CEFR was introduced in 2001, its least implemented concepts have later been elaborated on its companion volume (Council of Europe 2020). These include various aspects of learner agency, action-orientation in language teaching, mediation, and plurilingualism. The purpose of the new descriptors was to help language teachers and developers of the curricula to broaden the scope of language education and to value the linguistic and cultural diversity of learners. The CEFR companion volume describes (1) plurilingual and pluricultural competence and (2) communicative language competencies to help increase inclusiveness in language education. The former are based on building on the learners' pluricultural repertoire and plurilingual comprehension that scale up according to the learning level. The goal for C2, which is the highest level, is that the learner would be able to 'initiate and control their actions and forms of expression according to context, showing awareness of cultural differences and making subtle adjustments to prevent and/or repair misunderstandings and cultural incidents' (Council of Europe 2020, Chapter 4). This competence, among many other lingual and cultural competencies, transfers nicely into the intercultural sustainability competence of being able to communicate about complex matters of sustainable development in variable cultural and social contexts.

Communicative language competence is divided into linguistic competence (including, e.g., phonological control, grammatical accuracy, and vocabulary range), sociolinguistic competence (comprised of sociolinguistic appropriateness) and pragmatic competence (including, e.g., fluency, coherence, cohesion, and flexibility). An example of a readily transferrable competence can be found from the C2 level of developing sociolinguistic appropriateness: The aim is that the learner 'can effectively employ, both orally and in writing, a wide variety of sophisticated language to command, argue, persuade, dissuade, negotiate and counsel' (Council of Europe 2020, Chapter 5). Also here, the skills described can be used, e.g., in intergovernmental negotiations attempting to solve global environmental problems. This functions as another example where language-related competencies complement the sustainability competencies mentioned in the GreenComp framework (European Commission 2022). Conclusively, it can be stated that there are multiple competencies offered by language education that can complement and strengthen competencies needed in a sustainable future. It is more of a question of how language education and language teachers could connect the dots and enable a more comprehensive approach to language learning. This issue will be considered via the concepts of critical literacy of sustainable development and epistemic justice.

Critical Literacy of Sustainable Development and Epistemic Justice

Sustainable development as a policy includes a plethora of terms that can be ambiguous even to the native user of the North-Western global languages, e.g., English and French, which have been used when formulating its key terminology and concepts, such as sustainability, not to mention people with some other home language (Litre et al. 2022). While it can be argued that sustainable development in itself is a conflicting concept (see, e.g., Agbedahin 2019) and many researchers strive to steer away from its anthropocentric agenda towards a more ecocentric approach of thinking and living (McDonald 2014), it is still an ongoing and influential movement whose understanding is highly important to everyone, regardless of one's personal positioning.

Educators have a central role in driving societal changes (Bonner et al. 2020; Bürgener & Barth 2018; Weinberg et al. 2020; Koskela & Kärkkäinen 2021) and teachers in various subjects in the language educational field should logically be the starting point of building sustainable development literacy. As Agbedahin (2019) declares in their extensive literature review on sustainable development, ESD and UNESCO's Agenda 2030 Action Plan, educators need to understand the concepts involved in sustainable development and have a vision for ESD for being able (and willing) to adopt it in their teaching and curriculum planning. For learners to acquire literacy in the concepts concerning sustainable development, they need to have knowledgeable and motivated teachers who are capable of critical and systemic thinking. Hence, from the viewpoint of both learners and policymakers, increasing the sustainable development literacy of teachers is essential. Maijala and others (2024) suggested a didactic model for engaging language teachers in ESD through three simple phases: the teacher identifying the existing sustainability content in their language lessons, raising awareness by explaining to learners how the content and methods of ESD link to language learning, and finally, integrating ESD more systematically into language lessons. Further, to increase the sustainability literacy of teachers, instead of accepting the underlying neoliberal ideologies and North-Western conceptualisations, research by Misiaszek (2020) shows the benefits of using an ecopedagogical approach.

Ecopedagogy uses strategies of discerning ideological political agendas from their real-life consequences, i.e., deconstructing the concepts, especially that of *development*, to deepen and widen learners' understanding of the global wicked problems. In addition, Misiaszek (2020) argues that educators must be able to question the current framings of sustainable development and to have a dialogue with the learners about how to counter the gaps between ideologies and actualised social or environmental violence and unsustainable practices. Hence, instead of just learning the currently used terminology and concepts determined by the sustainable development policies, critical sustainable development literacy means that learners will learn how to look beyond the concepts, rethink their values and perceptions, and are enabled to discover new solutions to emerging, real-life environmental problems. Because of its dialogical approach and critical viewpoint, ecopedagogy seems an ideal tool for ESD in language teaching: it allows language teachers to use their preferred communicative methods to develop learner competencies without needing to understand everything about global environmental problems or even to agree with all aspects of sustainable development. Besides competencies needed in the changing world, language learners would benefit from increased relevance and quality of their education (Lupele and Lotz-Sisitka 2014), induced by new learning processes.

Many language teaching contexts are predominantly monolingual or based on few dominant language ideologies, which do not often include or even acknowledge multilingualism even when it is prevalent in the learner community (see, e.g., Rueda and Stillman 2012). Moreover, issues of sustainable development are primarily communicated in the language of education, such as English (Bexell and Jönsson 2021). As the main goal of ESD is to achieve the SDGs (UNESCO 2018), and their full understanding and local adaptation requires the use of local languages, there is strong evidence supporting a language-based approach as proposed by Mweri (2020) in their article on achieving the SDGs in the Global South context. This view is shared by Chapman (2022), who highlights the importance of Indigenous languages in increasing awareness of the SDGs outside academia. Consequently, the problematics of epistemic justice should be included when ESD is incorporated as part of language education. Mahboob (2023, p. 234) described epistemic injustice in education as 'excluding and/or silencing students' ways of knowing, doing, and being' and a related concept of linguicism as 'excluding and/or silencing students' other languages' in their open peer commentary on how to create a more inclusive educational environment.

Some of the recent advances in promoting epistemic justice include practices such as counterstorytelling to empower those subjected to epistemic violence (Dutta et al. 2022) and participatory research which aims at deliberating the capabilities of those who are or have been oppressed (Walker and Boni 2020). The work towards epistemic justice in language education is, unfortunately, still in its early stages as Kerfoot and Bello-Nonjengele (2023) demonstrate by stating that even advanced multilingual language teaching practices such as translanguaging (e.g., Creese and Blackledge 2010) fall short in epistemic justice because they can normally only offer epistemic access without changing the hierarchy of values or knowing relations. What is needed, instead, are inclusive practices that value all languages and their linguistic varieties and stem from a multilingual episteme.

Synthesis

This chapter aimed at linking language learning and ESD from the viewpoint of the learner and demonstrated how ESD can support language learning by offering meaningful learning contexts, increasing the quality of learning through effective learning processes and giving societal meaning to language learning (Lupele and Lotz-Sisitka 2014). The transformative ESD goals can, on the other hand, be supported by the modern view of language as a complex and adaptive system (Beckner et al. 2009), which the ongoing paradigm shift in language education enhances by increasing learner agency and active participation (Larsen-Freeman 2019). By embracing the emerging approaches that highlight different aspects of the same problematics related to transformative learning (Leaver et al. 2021), one can combine and make practical use of the research efforts in multiple fields of study (Oxford 2021). Competencies developed while learning languages were shown to coincide with many sustainability competencies. Especially competencies that develop skills of communication, pluricultural knowledge, and plurilingual comprehension (Council of Europe 2020) were applicable and supportive of the sustainability competencies in the GreenComp framework of the European Commission (2022).

In many Global South contexts, and increasingly also elsewhere, issues of epistemic injustice (Mahboob 2023) must be addressed to enable equity not only in all matters of sustainability (Litre et al. 2022) but in language learning contexts as well. This requires implementing ESD more holistically and systematically in language education and developing critical sustainable literacy in language teachers with the help of recent theories (Papi and Hiver 2024), didactic models (e.g., Maijala et al. 2024), practical implementations (e.g., Melin 2019; de la Fuente 2021), inclusive practices (e.g., Dutta et al. 2022), and pedagogical methods, such as ecopedagogy (Misiaszek 2020), that broaden the understanding of complex environmental problems and enable learners to discover solutions.

References

Agbedahin, A.V. 2019. Sustainable development, Education for Sustainable Development, and the 2030 Agenda for sustainable development: Emergence, efficacy, eminence, and future. *Sustainable Development* 27 (4): 669–680. https://doi.org/10.1002/sd.1931.

Baumgartner, L.M. 2012. Mezirow's theory of transformative learning from 1975 to present. In *The handbook of transformative learning*, ed. E.W. Taylor and P. Cranton. San Francisco: Jossey-Bass.

Beckner, C., R. Blythe, J. Bybee, M.H. Christiansen, W. Croft, N.C. Ellis, J. Holland, J. Ke, D. Larsen-Freeman, and T. Schoenemann. 2009. Language is a complex adaptive system: Position paper. *Language Learning* 59 (s1) (December): 1–26.

Bexell, M., and K. Jönsson. 2021. *The politics of the Sustainable Development Goals: Legitimacy, responsibility, and accountability*. Routledge studies in sustainable development. Place of publication not identified: Taylor & Francis.

Bonner, S. M., K. Diehl, and R. Trachtman. 2020. Teacher belief and agency development in bringing change to scale. *Journal of Educational Change* 21 (2) (May): 363–384. https://doi.org/10.1007/s10833-019-09360-4.

Bürgener, L., and M. Barth. 2018. Sustainability competencies in teacher education: Making teacher education count in everyday school practice. *Journal of Cleaner Production,* 174 (February): 821–826. https://doi.org/10.1016/j.jclepro.2017.10.263.

Canning, J. 2010. Translating words into action and action into words: Sustainability in languages, linguistics and area studies curricula. In *Sustainability education: Perspectives and practice across higher education,* ed. P. Jones, D. Selby, and S. Sterling, 295–312. London: Routledge.

Chapman, R.S. 2022. Sustaining languages and the language of sustainability. *Iperstoria* 20: 78–98. https://doi.org/10.13136/2281-4582/2022.I20.1184.

Council of Europe. 2001. *Common European framework of reference for languages: Learning, teaching, assessment*. Strasbourg: Council of Europe. Available from: https://rm.coe.int/1680459f97 (accessed August 6, 2024).

Council of Europe. 2020. *Common European framework of reference for languages: Learning, teaching, assessment – Companion volume*. Council of Europe Publishing, Strasbourg. Available from: www.coe.int/lang-cefr. (accessed August 6, 2024).

Creese, A., and P. Martin. 2003. Multilingual classroom ecologies: Interrelationships, interactions and ideologies. *International Journal of Bilingual Education and Bilingualism* 6 (3–4) (June): 161–167. https://doi.org/10.1080/13670050308667778.

Creese, A., and A. Blackledge. 2010. Translanguaging in the bilingual classroom: A pedagogy for learning and teaching? *The Modern Language Journal* 94 (1) (March): 103–115. https://doi.org/10.1111/j.1540-4781.2009.00986.x.

Dirkx, J.M. 2001. The power of feelings: Emotion, imagination, and the construction of meaning in adult learning. In *The new update on adult learning theory. New directions for adult and continuing education,* vol. 89, ed. S.B. Merriam, 63–72. SanFrancisco, CA: Jossey-Bass.

Dutta, U., A.K. Azad, and S.M. Hussain. 2022. Counterstorytelling as epistemic justice: Decolonial community-based praxis from the Global South. *American Journal of Community Psychology* 69 (1–2) (March): 59–70. https://doi.org/10.1002/ajcp.12545.

European Commission. Joint Research Centre. 2022. *GreenComp, the European Sustainability Competence Framework.* LU: Publications Office. Available from: https://doi.org/10.2760/13286. (accessed August 6, 2024).

de la Fuente, M.J., ed. 2021. *Education for Sustainable Development in foreign language learning: Content-based instruction in college-level curricula*, 1st ed. Routledge, Research in Language Education. New York: Routledge.

García, O., and L. Wei. 2015. Translanguaging, bilingualism, and bilingual education. In *The handbook of bilingual and multilingual education*, ed. W.E. Wright, S. Boun, and O. García, 223–240, 1st ed. Wiley. https://doi.org/10.1002/9781118533406.ch13.

Halliday, M.A.K. 1993. Towards a language-based theory of learning. *Linguistics and Education*, 5: 93–116.

Hammadou, J., and E.B. Bernhardt. 1987. On being and becoming a foreign language teacher. *Theory into Practice* 26 (4) (September): 301–306. https://doi.org/10.1080/00405848709543290.

Kerfoot, C., and B.O. Bello-Nonjengele. 2023. Towards epistemic justice: Constructing knowers in multilingual classrooms. *Applied Linguistics* 44 (3) (July 1): 462–484. https://doi.org/10.1093/applin/amac049.

Koskela, T., and S. Kärkkäinen. 2021. Student teachers' change agency in Education for Sustainable Development. *Journal of Teacher Education for Sustainability* 23 (1) (June 1): 84–98. https://doi.org/10.2478/jtes-2021-0007.

Kramsch, C. 2014. Teaching foreign languages in an era of globalization: Introduction. *The Modern Language Journal* 98 (1): 296–311.

Larsen-Freeman, D. 2019. On language learner agency: A complex dynamic systems theory perspective. *The Modern Language Journal* 103: 61–79.

Leaver, B.L. 2021. Transformative language learning and teaching: The next paradigm shift and its historical context. In *Transformative language learning and teaching*, ed. B.L. Leaver, D.E. Davidson, and C. Campbell, 1st ed. Cambridge: Cambridge University Press. Available from: https://www.cambridge.org/core/product/identifier/9781108870788/type/book (accessed August 6, 2024).

Leaver, B.L., D.E. Davidson, and C. Campbell, ed. 2021. *Transformative language learning and teaching*, 1st ed. Cambridge University Press. https://doi.org/10.1017/9781108870788.

Léglise, I. 2017. Multilinguisme et hétérogénéité des pratiques langagières. Nouveaux chantiers et enjeux du Global South. [Multilingualism and heterogeneous language practices: New research areas and issues in the

Global South.] *Langage et Société* N° 160–161 (2) (May 17): 251–266. https://doi.org/10.3917/ls.160.0251.

Liddicoat, A.J., and A. Scarino. 2013. *Intercultural language teaching and learning*. Oxford: Blackwell Publishing Ltd. https://doi.org/10.1002/978 1118482070.

Litre, G., F. Hirsch, P. Caron, A. Andrason, N. Bonnardel, V. Fointiat, W.O. Nekoto, J. Abbot, C. Dobre, J. Dalboni, A. Steuckardt, G. Luxardo, and H. Bohbot. 2022. Participatory detection of language barriers towards multilingual sustainability(ies) in Africa. *Sustainability* 14 (13) (July 4): 8133. https://doi.org/10.3390/su14138133.

Lupele, J., and H. Lotz-Sisitka. 2014. *Learning today for tomorrow: Sustainable development learning processes in sub-Saharan Africa*. SADC REEP/ Share-Net. https://doi.org/10.13140/2.1.3559.2643.

Mahboob, A. 2023. On linguicism, epistemic injustice, and research in language-in-education: A commentary on 'midadolescents" language learning at school: Toward more just and scientifically rigorous practices in research and education. *Language Learning* 73 (S2): 234–238. https://doi.org/10.1111/lang.12570.

Maijala, M., N. Gericke, S.-R. Kuusalu, L.M. Heikkola, M. Mutta, K. Mäntylä, and J. Rose. 2024. Conceptualising transformative language teaching for sustainability and why it is needed. *Environmental Education Research* 1–20. https://doi.org/10.1080/13504622.2023.2167941.

Makalela, L. 2017. Bilingualism in South Africa: Reconnecting with Ubuntu translanguaging. In *Bilingual and multilingual education*, ed. O. García, A. M. Y. Lin, and S. May. Cham: Springer International Publishing. Available from: https://doi.org/10.1007/978-3-319-02258-1_14 (accessed July 12, 2024).

McDonald, H.P. 2014. *Environmental philosophy: A revaluation of cosmopolitan ethics from an ecocentric standpoint,* BRILL, ProQuest Ebook Central. Available from: https://ebookcentral.proquest.com/lib/kutu/detail.action?docID=1686960 (accessed July 12, 2024).

Melin, C., Ed. 2019. Foreign language teaching and the environment: Theory, curricula, institutional structures. *Teaching Languages, Literature, and Culture*, 6. New York: The Modern Language Association of America.

Mezirow, J. 1978. Perspective transformation. *Adult Education* 28 (2): 100–110. https://doi.org/10.1177/074171367802800202.

Misiaszek, G.W. 2020. Ecopedagogy: Teaching critical literacies of 'development', 'sustainability', and 'sustainable development'. *Teaching in Higher*

Education 25 (5) (July 3): 615–632. https://doi.org/10.1080/13562517.2019.1586668.
Mukute, M., and H. Lotz-Sisitka. 2012. Working with cultural-historical activity theory and critical realism to investigate and expand farmer learning in Southern Africa. *Mind, Culture, and Activity* 19 (4) (October): 342–367. https://doi.org/10.1080/10749039.2012.656173.
Mweri, J. 2020. Sustainable Development Goals: Reaching people through their mother tongue. *Linguistics and Literature Studies* 8 (1) (January): 14–25. https://doi.org/10.13189/lls.2020.080103
Norro, S. 2021. Namibian teachers' beliefs about medium of instruction and language education policy implementation. *Language Matters* 52 (3) (September 2): 45–71. https://doi.org/10.1080/10228195.2021.1951334.
Oxford R.L. 2021. Shaking the foundations: Transformative learning in the field of teaching English to speakers of other languages. In *Transformative language learning and teaching*, ed. B.L. Leaver, D.E. Davidson, and C. Campbell. Cambridge University Press.
Papi, M., and P. Hiver. 2024. Proactive language learning theory. *Language Learning*. https://doi.org/10.1111/lang.12644.
Rueda, R., and J. Stillman. 2012. The 21st century teacher: A cultural perspective. *Journal of Teacher Education* 63 (4) (September): 245–253. https://doi.org/10.1177/0022487112446511.
Sinakou, E., V. Donche, J. Boeve-de Pauw, and P. Van Petegem. 2019. Designing powerful learning environments in Education for Sustainable Development: A conceptual framework. *Sustainability* 11 (21) (October 28): 5994. https://doi.org/10.3390/su11215994.
Stibbe, A. 2015. *Ecolinguistics: Language, ecology and the stories we live by*. London; New York: Routledge, Taylor & Francis Group.
Stibbe, A. 2024. *Econarrative: Ethics, ecology, and the search for new narratives to live by*. London, New York, Oxford, New Delhi, Sydney: Bloomsbury Academic.
Tikly, L. 2019. *Education for Sustainable Development in the postcolonial world: Towards a transformative agenda for Africa*, 1st ed. Routledge. https://www.taylorfrancis.com/books/9781351812405.
Tilbury, D. 2011. *Education for Sustainable Development: An expert review of processes and learning. France,* UNESCO. Available from: https://unesdoc.unesco.org/ark:/48223/pf0000191442 (accessed July 12, 2024).
UNESCO. 2018. *Global Action Programme on Education for Sustainable Development. Education for Sustainable Development and the SDGs. Policy brief:*

Advancing ESD policy. France, United Nations Educational, Scientific and Cultural Organization.

Walker, M., and A. Boni. 2020. Epistemic justice, participatory research and valuable capabilities. In *Participatory research, capabilities and epistemic justice*, ed. M. Walker and A. Boni. Cham: Springer International Publishing. https://doi.org/10.1007/978-3-030-56197-0_1.

Vare, P., and W. Scott. 2007. Learning for a change: Exploring the relationship between education and sustainable development. *Journal of Education for Sustainable Development* 1 (2) (September): 191–198. https://doi.org/10.1177/097340820700100209.

Weinberg, A.E., C.D. Trott, W. Wakefield, E.G. Merritt, and L. Archambault. 2020. Looking inward, outward, and forward: Exploring the process of transformative learning in teacher education for a sustainable future. *Sustainability Science* 15 (6) (November): 1767–1787. https://doi.org/10.1007/s11625-020-00831-9.

Virtanen, P.K., L. Siragusa, and H. Guttorm. 2020. Editorial overview: Indigenous conceptualizations of 'sustainability.' *Current Opinion in Environmental Sustainability* 43 (April): A1–A2. https://doi.org/10.1016/j.cosust.2020.04.004.

4

Unsustainability and Ecolinguistics

Attila Krizsán

Introduction: Applied Linguistics and Sustainable Futures

This chapter discusses the current dominant conceptualisation of sustainability and some of its alternatives. Furthermore, it proposes a way in which linguistic research could contribute to new conceptualisations of sustainability. I argue that the impact of ecologically oriented linguistic research on the notion of sustainability (1) could advance new ways of language use and (2) support change towards more sustainable relationships with the ecosphere. I briefly introduce current linguistic scholarship and will limit my scope to a specific area: the role of applied linguistic research—especially functional-, critical- and ecolinguistics—in reconceptualising and changing sustainability-related communication, and, through this, thinking and action.

A. Krizsán (✉)
School of Languages and Translation Studies, University of Turku, Turku, Finland
e-mail: atarkr@utu.fi

M. Maijala et al. (eds.), *Transformative Language Teaching for Sustainability*,
https://doi.org/10.1007/978-3-031-85493-4_4

In the last two decades, there have been a number of applied linguistic studies on topics related to sustainability. Nervino et al. (2024) conducted a bibliometric study and recorded 114 relevant articles starting from 2004. In terms of the frameworks used in these papers, the authors (pp. 876–877) listed corpus linguistics as the most frequently occurring, which, until recently, was often used in combination with critical discourse analysis and discourse analysis. However, according to Nervino and others (2024), ecolinguistics was not applied as much as expected, not even in studies on environmental sustainability. In addition, while biocentrism was mentioned in connection with topics in communication sustainability studies, it was lacking in studies related to language and linguistics. Moreover, alternative views of sustainability related to, for example, non-anthropocentrism, sentiocentrism (i.e., a moral-philosophical approach based on the considerability of all sentient beings), holism and regenerative sustainability also appeared to be lacking in the studies examined.

There was a growing focus on multimodality in the papers, which aligns with our current ways of, for example, online multimodal interaction. However, it seemed that multimodality in these studies was limited to the usage of multiple modes (e.g., images and text) in, for example, media communication. Other modes that are crucial for understanding humans and their relationships with their environment, such as designs of human spaces (including spaces for non-human animals), were absent from these studies. Against this backdrop, the goal of the present chapter is to broaden the perspectives of applied linguistic studies on sustainability. This is done by (1) critiquing the concept of sustainability and outlining new directions by emphasising the need for non-anthropocentric understandings of the concept, (2) presenting a selection of non-anthropocentric approaches to sustainability and (3) advancing ideas related to an ecolinguistic analytic approach to semiotic modes in spatial design and architecture that could reveal further alternatives.

Noteworthy, applied linguistic research is not unique among fields of science in its potential to trigger change towards a more sustainable world. In fact, numerous fields have been examining conceptualisations of sustainability with the purpose of finding ways to rethink the roles of human beings in the ecosphere. Several fields have advanced new understandings of the relationships between humans and other elements of nature. In this chapter, I will also draw on perspectives from two fields: (1) architecture and landscape sustainability and (2) comparative education. These two areas have been at the forefront of novel proposals regarding ways of sustainable thinking and action in terms of how humans relate to their living habitat and how we reconstruct cultural forms of knowledge through education. Both of these aspects of human life and forms of knowledge relate to sustainability. Furthermore, they are relevant from an (eco)linguistic point of view due to the central role of language use/semiosis in understanding and relating to human habitat and cultural practices.

Despite not being the only field with a considerable potential for change, the role of applied linguistics in sustainable futures is notable because the information we receive, consume and reproduce through various means of social semiosis (i.e., language use performed through varied media and in multiple contexts) plays a significant role in our understandings of the world, including how we position ourselves in it. These understandings have a substantial influence on the identities available for us to perform, our ways of belonging, legitimisation for our actions and the justifications for how we live (see e.g., van Dijk 2016). Thus, language use is at the core of how we relate in practice to ourselves, other human beings, other-than-human beings and the non-living parts of our environment. In light of the immense influence of humanity on our ecosystems, we are responsible for understanding and changing the ways in which we relate to human and non-human others and the environment to assure a healthy ecosphere for all of its components. Alternatively, according to Rappaport (1999, p. 461), 'if evolution, human and otherwise is to continue, humanity must think, not only about the world, but on behalf of the world of which it has become a special part and to which, therefore [...] it has enormous responsibilities'.

However, to think on behalf of the world (or, rather, with the world), we have to be able to reflect on our roles within the ecosphere and understand the connections of our cultures and ways of being in relation to these roles. We also need to understand and assist human life in becoming a more sustainable part of this ecosphere as a whole. Employing a focus on ecological fairness, applied linguistic research practices directed at our norms and semiotic representations of our ways of being could be one of the first steps towards such understanding. Furthermore, this understanding could lay the groundwork for thinking about and communicating with the world as well as assist in overcoming our current global *culture of unsustainability* (Syrjämaa et al. 2021). In practice, such change could be driven by discovering, through linguistic research, new and renewed values that inform our roles in the world. Through these discoveries, we could find ways of language use and teaching practices directed at raising bioecological awareness (see Steffensen et al. 2024) and overcoming the influence of human exceptionalism (see e.g., Silova 2021).

I elaborate on these ideas in greater detail in the sections below. In section 'The Anthropocene as a time of crises', I focus on unsustainability and provide a brief insight into the climate crisis and its relation to human exceptionalism and our current dominant approach to sustainability, that is, sustainable development. In section 'Alternative views on sustainability', I centre the discussion of the topic mostly on environmental sustainability, which is interconnected with cultural, social, economic and security-related sustainability. Looking beyond human exceptionalism and sustainable development through the lens of (eco)critical linguistic research is relevant for analysing communication related to sustainability and for advancing new ways of sustainability-related communication and thinking, that is, new econarratives (see Stibbe 2024). In section 'Maintaining unsustainability through language', I move on to powerful institutional texts and narratives that contribute to maintaining unsustainability. In section 'Targeting unsustainability through multimodal ecolinguistic research', I discuss the role of language use/semiosis and focus on studying these factors from a (multimodal) ecolinguistic perspective. The final section summarises and concludes the chapter.

The Anthropocene as a Time of Crises: Human Exceptionalism and the Development Paradigm of Sustainability

Iveta Silova, the president of the Comparative and International Education Society, described the Anthropocene—the period of human-induced ecological crises on Earth—as 'a process that amplifies death to such extent that not only individuals, but entire groups, species, and ways of life are being eradicated, one involving the death of future continuities' (Silova 2021, p. 587). Indeed, the ecological crisis is severe, even when the perspective is limited to the needs of a single Earthly species—humans. According to the United Nations World Water Development Report, by 2030, there will be a 40 per cent global shortfall between the available water supply and the predicted demand for water (UNESCO 2024a) and that by 2050, 90 per cent of the land surface of Earth could be degraded (UNESCO 2024b). Both of these scenarios pose enormous threats to biodiversity and human life, especially the lives of people whose access to water and land area is characterised by scarcity. However, the loss is much more severe if we think beyond humanity and consider the lives of other-than-human beings who have even fewer crisis management tools at their disposal.

Besides the fact that large-scale industrial farming causes extensive suffering to non-human animals, it also has severe consequences for humanity. In fact, we should consider other-than-human beings even if our goal solely involved 'meeting the needs of the present without compromising the ability of future [human] generations to meet their own needs', according to the prevailing definition of sustainability (Brundtland 1987). This is because the fates of human and other-than-human beings are closely intertwined, perhaps more than ever before, due to the enormous impact of the industrial capitalist societies of the Global North on the ecosphere. The interconnectedness of the lives of non-human beings with the dominant human ways of life is illustrated by the consequences of animal farming. It is well known that industrial farming contributes to deforestation, freshwater shortage, soil erosion and biodiversity loss (Jalava et al. 2024; JYU.Wisdom community 2021).

Due to the impact of industrial farming, the global proportion of the mass of wild mammals versus other mammals is four versus 96 per cent. Of the 96 per cent of all mammals who do not count as wild animals, 36 per cent account for humans and 60 per cent for farmed animals. With regard to birds (who account for high usage in the animal industry), the percentages of wild and farmed birds are 30 versus 70 per cent, respectively. In terms of the global mass of non-human animals, the proportion of domesticated animals is tenfold that of wild animals, or 620 million tonnes in total (Bar-On et al. 2018).

Furthermore, the role of industrial animal farming is questioned even within the paradigm of *resource efficiency* or the *food economy* as well as in relation to the second United Nations' (2015) Sustainable Development Goal (SDG): zero hunger. According to Smithers (2017), the amount of food fed to farm animals for meat production is more than enough to feed the entire human population in need of food. Although Smithers' report only talks about food quantities, it makes a strong argument for the current ways of industrial farming, which contribute to short and difficult lives for farmed animals and serve the wealthy at the expense of the poor within and across human societies. Furthermore, animal exploitation leads to interconnected problems related to global inequality, as various types of oppression, such as racism, ableism, sexism and classism (see e.g., Martusewicz 2018), go hand in hand with our taken-for-granted idea of human supremacy. Alternatively, as the linguist M.A.K. Halliday put it more than twenty years ago, 'the hegemony of human species is inseparable from the hegemony usurped by one human group over another, and that neither will come to an end as long as the other still prevails' (Halliday 2001, pp. 198–199). Halliday's thoughts are also echoed in Crist (2019), who saw human exceptionalism—that is, a belief in the superiority of humans over other life forms and our right to use other-than-human beings and their habitat for our own purposes—as the main cause of the climate crisis.

Driven by human exceptionalism, consumerist lifestyles in the Global North and areas influenced by its promotional culture are built on the prevalence of dominant anthropocentric, androcentric and white-supremacy-based identities. One example of the intersection of these identities in a lifestyle is visible through so-called *petro-machoism*. Petro-machoism illustrates the connection between white patriarchy and fossil fuels (Caimotto 2024) and responds to a culturally generated demand built on aspects of identities related to manhood and power. In addition to connecting misogyny and climate change denial, petro-machoism showcases 'an attachment to the righteousness of fossil fuel lifestyles, and to all the hierarchies that depend upon fossil fuel, [and] produces a desire to not just deny, but to refuse climate change' (Daggett 2018, p. 41). Therefore, climate change denial is not simply a problem of individuals having harmful values; it is also a byproduct of the narratives of industrial capitalism in the world, including those regarding the roles we play. The climate crisis is but a clear sign that the globalised desirable ways of life for those who can afford to strive for them are built on Global North narratives of the *good life*. These ways of life are also fuelled by industrial overproduction that require an ever-growing number of resources and are geared towards promoting unsustainable lifestyles by default due to the emphasis on permanent growth and progress. In societies characterised by the attention economy of the Global North, industrial narratives are constantly communicated in real places and virtual spaces, where—to recycle a term used by Althusser (1971)—they *interpellate* us through layered mixtures of lifestyles made available and desirable via semiotic means. The products of our economies, evident in the case of cars, also have semiotic functions in communicating and realising particular lifestyles and, from an ecocritical point of view, can also be studied with regard to their role in communicating narratives and offering identity positions. As I will propose in section 'Targeting unsustainability through multimodal ecolinguistic research', combining the present state of ecolinguistic research with multimodal analytic methods could advance the investigation of these material domains in terms of narrative production.

The influence of narratives of permanent growth—leading to a dependency of ever-increasing cycles of production and consumption—can also be traced in the United Nations SDGs, which cement the political obsession with economic growth, fail to challenge the overconsumption of the rich and are almost entirely anthropocentric (e.g., Stibbe 2021). This is not surprising, as the current dominant understanding of sustainability is geared towards supporting industrial capitalism and, along with it, unsustainable lifestyles in the Global North. Hickel (2015) criticised the SDGs, maintaining that they 'are actively dangerous: they lock in the global development agenda for the next 15 years around a failing economic model that requires urgent and deep structural changes [...] [T]he core of the SDG programme for development and poverty reduction relies precisely on the old model of industrial growth—ever-increasing levels of extraction, production, and consumption' (Hickel 2015 in Stibbe 2019, p. 237). Stibbe (2019) went on to describe the ways in which several of the SDGs contribute to maintaining the unsustainable growth paradigm, for example, SDG12, which is based on resource efficiency, industrialisation and economic growth; SDG 14, which refers to sea-life and seascapes as marine resources, and SDG15, which advances the idea of genetic resource use. Virtanen et al. (2020) identified a similar economic growth-focused Western approach in the SDGs. They maintained that the SDGs are motivated by the ideas of ownership and personhood and fail to consider the broader global context, for example, the needs of Indigenous communities.

In addition to being built on unsustainable economic principles, several studies have criticised the SDGs for their anthropocentricity. For example, Torpman and Röcklingsberg (2021) expressed concerns related to the lack of direct consideration of non-human animals in the SDGs. They see this as a problem for (1) instrumental reasons, that is, related to the benefit of human beings, such as avoiding risks related to diseases spreading due to consuming animals (e.g., in the case of COVID-19) and the increase in bacterial resistance to antibiotics due to modern animal farming, and (2) reasons grounded in the intrinsic value of non-human animals, that is, valuing the lives of non-human animals in and of themselves.

Alternative Views on Sustainability

In addition to their criticism, Torpman and Röcklingsberg (2021) also proposed an approach which alters the current SDGs in a way that enables the inclusion of the needs of non-human animals in SDGs two, three, six, 10, 11, 12, 13, 14 and 15. Approaching the problem from the perspective of sentientism—that is, a moral-ethical perspective that recognises the intrinsic value of beings that are able to experience pleasure and pain—the human-centred SDGs that do not include the needs of non-human animals can also be viewed as sentience-centred, as humans are also sentient beings. Other researchers (e.g., Visseren-Hamakers 2020) are also seeking solutions from a sentientist stance. Visseren-Hamakers (2020) proposed an 18th SDG that takes into consideration the evolving values of human–nature relationships exemplified in other official texts. These include 'the recognition of animal rights in the German constitution; recognition of rights of nature, e.g., in Ecuador, Colombia, India, New Zealand, and communities in the United States; granting rights to rivers, e.g., the Whanganui river in New Zealand and the Ganges and Yamuna rivers in India; The United Nations World Charter for Nature; The Universal Declaration of Rights of Mother Earth; The Earth Charter; and The World Declaration on Great Apes' (Visseren-Hamakers 2020, p. 1). From the moral philosophical position of sentientism, anthropocentric concepts of sustainability need to be extended due to the fact that sustainable development, in its current form, is characterised by 'exclusive validity for human beings and its reduction of non-human animals to human capital' (Boscardin and Bossert 2015, p. 2). On top of anthropocentrism, traces of the principles of economic supply and demand within SDGs can also be criticised. These principles, which form part of the paradigm of neoclassical economics, de facto ignore demand without financial backing (i.e., effective demand), such as the case regarding the need for sanitation systems in areas where inhabitants cannot afford them (see Chau et al. 2022). Therefore, from a sentientist perspective, widening the scope of the SDGs to include the needs of all sentient beings could be grounded on similar moral principles as in the case of humans who are unable to represent their demands through the economy but are considered in, for example, SDGs one, two and six.

In addition to sentientist approaches to sustainability, the anthropocentrism of the paradigm of sustainable development has also been contested by holistic perspectives on the concept. One particularly interesting holistic approach is that of regenerative sustainability. In the field of architecture and landscape sustainability, Gibbons (2020) presented a detailed comparison of conventional, contemporary and regenerative sustainability and assembled an extensive list of regenerative development principles (pp. 9–14). In a comparison of conventional sustainability (i.e., sustainability as conservation), the contemporary sustainable development paradigm (i.e., sustainability as ecosystem services to humans) and regenerative sustainability (a holistic view on sustainability that sees human and other lives as part of an autopoietic system, i.e., a system maintained by self-creating processes), Regenerative Development Goals (RDGs) are seen as elevated SDGs. In contrast to SDGs, RDGs emphasise processes alongside outcomes. These processes involve engagement between places, and both human and non-human inhabitants with the goal of regeneration, and are based on a list of principles, such as wholeness (i.e., processes work in whole systems) and change (i.e., manifesting potential in a place rather than solving problems).

Gibbons prioritised the move towards a more holistic approach and criticised sustainable development for its inability to move beyond its grounding on the mechanistic worldview of conventional sustainability and for being reductionist, fragmented and reliant on efficiency and technological fixes. According to Gibbons (2020, p. 19), '[c]ontemporary sustainability mostly focuses on symptoms rather than causes of unsustainability—in other words, shallow leverage points in systems, such as technological, policy, and economic changes—and supports the continued existence of unsustainable systems and thought patterns (e.g., continued economic growth, managing environmental resources for human consumption, efficiency)'. In comparison to the mainstream understanding of sustainability, a regenerative approach is based on ideas of humans and other elements of nature as a whole and supports actions to regenerate ecosystems as opposed to simply maintaining them or causing them net-zero harm. Regenerative sustainability aims to reinforce 'the health, adaptive capacity, and evolutionary potential of the fully integrated global social-ecological system so that it can continue regenerating itself, thereby creating the conditions for a thriving and abundant future—not only for the human species but for all life' (DuPlessis and Brandon 2015, p. 59).

Comparable ideas of holism can also be traced in the planetary well-being approach by the Wisdom community of the University of

Jyväskylä, Finland (JYU.Wisdom community 2021). In their systems-oriented framework, the authors see humanity as part of nature and emphasise the relevance of the systems and processes required for satisfying the needs of the life forms on our shared planet. Systems, in this view, can be exemplified by organisms, populations, lineages, and ecosystems. These systems are dependent on critical system processes, which are required for each system to realise their system-specific characteristics (e.g., persistence). The emphasis in this approach is on so-called Earth system processes (e.g., the global climate) and ecosystem processes (e.g., succession). Planetary well-being is seen 'as a state in which the integrity of Earth system and ecosystem processes remains unimpaired to a degree that lineages can persist to the future as parts of ecosystems, and organisms (including humans) can realise their typical characteristics and capacities' (JYU.Wisdom community 2021, p. 4). The authors proposed that their approach be used mainly as a new perspective on the ecological crisis that can aid thinking, research and actions towards the sustainability transition beyond anthropocentrism.

Naturally, education has a key role in this transition. It can support the current situation by spreading narratives of human exceptionalism and eternal growth or drive change by seeking out and promoting alternative narratives. According to Rappleye and Komatsu (2020), current forms of education intensify the climate crisis, leading to social injustice; this is due to overarching paradigms of, for example, comparative education (e.g., increasing literacy and improvements in student scores) being based on the promotion of economic growth via the presumed connection between national economic growth and cognitive skills or student scores. National educational goals are often based on educational principles developed on the basis of, for example, shared *best practices* or ideas related to *personal growth and development*. These factors reflect prevailing ideas in the business world and echo principles that attract criticism of the industrialist capitalist thinking of the Global North (see e.g., Stibbe 2024). The result is that global inequality is spreading, poverty is increasing, and we are closer to an environmental breakdown than ever before, despite more than sixty years of international development efforts. Thus, the Global North, in general, does not seem to have understood the influence and role of education during the Anthropocene (see e.g., Silova 2021). Therefore, to live more sustainably, we should think, speak, and educate beyond the basic values behind Western industrial capitalism (see e.g., Plumwood 2001; Haraway 2016a). Education, thus, has enormous potential in terms of building on the interconnectedness of culture, ways of language use, and thinking.

On the basis of the above-presented alternative views on sustainability, I argue that approaches based on sharing and co-creating in a system consisting of humans and other elements of nature as a whole could better encompass the well-being of humans, non-human animals and the ecosystems on which life depends than the anthropocentric orientation of the prevailing sustainable development paradigm. In addition, replacing the progress-centred aspect of (sustainable) development with adaptation and cohabitation, the interdependence and relationality in the permanently changing and interacting complex systems of biotic and abiotic components that form nature (see Haraway 2016a) could inspire new ways of communication and encourage new ways of being via, e.g., promoting ecocultural identities (see Stibbe 2024). This shift can lead to fairer and healthier ecosystems, where sustainability is approached in terms of relationships and networks among all components, as opposed to artificially separated dimensions or pillars motivated only by human benefit (but do not in fact serve this benefit for the above-mentioned reasons) (see Haraway 2016a; Gibbons 2020).

The purpose of this section was to present an overview of alternative views regarding sustainability and the values they espouse. I shall turn to the ways in which our relationship with sustainability is grounded in language use and the potential of changing the ways we communicate for the shift towards more sustainable societies.

Maintaining Unsustainability Through Language: Powerful Texts and Narratives on Humans and Other Elements of Nature

According to Stibbe (2024), the way we use language, think and act is influenced by narratives. He maintains that narratives are cognitive structures that we repeatedly encounter in our cultures via semiotic means, for example, by learning about histories and being targeted by advertisements and entertained by films and novels (see Stibbe 2024). These ideas about the interconnectedness of semiosis, cognition, and society are also supported by the discourse-cognition-society triangle of van Dijk's (2016) socio-cognitive model. Furthermore, the dependency of a culture on narratives has also been examined by Ricoeur (1985). Ricoeur (1984), who saw narratives as arrangements of events, also underlined their centrality in communication, culture and human experience; he argued that they are the cornerstones of our realities in terms of identity, power and knowledge. Furthermore, Ricoeur proposed that through the intersection of the world of the recipient of a text and the world of a narrative, narratives are capable of transforming the ways in which recipients experience or perceive the world.

A look at semiosis beyond text and image shows that designs of human spaces also communicate socially relevant narratives via emphasising, backgrounding or omitting particular values or functions; we can think of, for example, socialist realist buildings with their toned-down colours, often simple arrangements of windows, art scenes depicting the lives of the working class and massive shapes to communicate power; we can then contrast these with the archetypical modern glass-covered skyscrapers in metropolitan areas signifying wealth, efficiency and technological advancement. Human influence on the world, like human communication, is embedded in shared cultural values. These values are reflected in what we do, how we do it, what narratives we communicate, and how and where we communicate them. The latter means that the contexts where narratives or the semiotic aspects of human activities are perceived are also part of their meaning-making potential.

In line with Fløttum and Gjerstad (2017) and Stibbe (2024), I see the roles of narratives as crucial in the communication of shared frames and claim that the majority of our current social and environmental crises are maintained through naturalised (on naturalisation, see Barthes 1972; Fairclough 1995) ways of thinking built via taken-for-granted narratives in the industrial societies of the Global North. Inspired by Lyotard's (1984) concept of grand narrative, Stibbe (2024, p. 18) viewed metanarrative as 'a general worldview that is held across the minds of multiple people within a culture as a cognitive structure and manifested in particular texts'. Lyotard (1984) claimed that metanarratives function as socially legitimised narratives behind, for example, narratives of experience, history and knowledge that shape the ways humans understand the world. Lyotard exemplified metanarratives by the metanarrative of *progress* behind economic growth, the metanarrative of *wealth creation* in liberal democracies, or the metanarrative of the *emancipation* of the working class in Marxism. According to him, metanarratives, such as those in the above examples, inform philosophies on the historical process. Thus, metanarratives influence human worldviews and actions by making particular positions—such as human-centred, male-centred or growth-centred approaches to our ecosystem in the case of *progress*—seem natural and historically grounded. However, metanarratives do not just influence our understanding of the world as something we are encultured into; they are constantly reproduced through semiotic means, including powerful institutional texts.

One example of a human-centred metanarrative is evident in the five pillars of sustainability ('People, Prosperity, Planet, Peace, and Partnership') in the 2030 Agenda for Sustainable Development (United Nations 2015). According to Tremblay et al. (2020, p. 12), 'the 2030 Agenda is based on the principles of universality, inclusion—reflected by the "leaving no one behind" motto—, and integrality'. However, non-human animals are not included among those that should not be left behind, not even in the pillar labelled 'Planet', where they only appear among marine resources and biodiversity loss. Thus, the UN Agenda 2030, in line with the Brundtland Report's definition of sustainability encountered earlier, reinforces the same metanarrative, one where the role of nature is to serve human need. Both the 2030 Agenda for Sustainable Development and the Brundtland Report are influential texts, not just in terms of the values they promote but also due to their institutional and legal power. The Brundtland Report, for example, still serves as the basis of EU law on sustainable development and lies at the heart of the approach taken by legal institutions on sustainability around Europe.

Somewhat surprisingly, an example of the prioritisation of economic growth over more sustainable goals regarding the ecosystem as a whole can also be found in texts aimed at promoting global sustainability-based thinking and action. For example, the conceptualisation of sustainability in the Summary for Policymakers of the Intergovernmental Panel on Climate Change (IPCC)—a mainstream international and supranational organisation tasked with assessing the science on climate change—is also framed, in part, by industrial, economy-centred thinking (see Kanerva and Krizsán 2021). Boscardin and Bossert (2015) discussed this problem by asserting that current IPCC reports represent the approach of ecological modernisation theory, which promotes technological and managerial tools for promoting a combination of environmental conservation and economic growth in a viable and profitable way, and reflect *weak sustainability*. In an interview conducted in Hungarian, Diána Ürge-Vorsatz, the vice-chair of the IPCC, expressed her personal views on sustainability by heavily criticising our current oil and gas-based economies and the growth-based narratives of industrial capitalism. Among other things related to the problematic feasibility of the current approach of cutting emissions, Ürge-Vorsatz expressed as follows:

> We have to stop using fossil fuels; we should not burn more oil and gas, although even doing so would not reverse the process [of climate change] and I think that we have to take the compulsion of permanent growth out of the economy because it does not only harm the environment, but it is also unhealthy for us […] we need to change the whole foundation of the system. It is useless to blame the companies, since they are under pressure too due to the expectation of constant and ruthless growth by the stock markets. (Ürge-Vorsatz 2023, translated by the author)

These thoughts and the above criticism on IPCC reports illustrate how strongly even institutions such as the IPCC are influenced by industrial capitalism. The ways the IPCC is designed, for example, the type of interest representation within the institution, undermines its capability to effect change on the basis of scientific findings. This is true regardless of the views of IPCC members who are in line with the findings. Therefore, institutions such as the IPCC need to have more liberty to be able to put science ahead of politics and freedom from attempting to comply with the demands of member states and the lobby groups involved in the work of the institution (see Kanerva and Krizsán 2021). For societal change to take place, there is an urgent need for sustainability-related multidisciplinary thinking, research, education, communication and activism. In the words of Ürge-Vorsatz 'we can only trigger the right individual action by a system-level change, but no change will happen until we reach the critical mass' (Ürge-Vorsatz 2023, translated by the author).

Due to the strong role of semiosis in maintaining cultures of unsustainability in several social domains (see section 'The Anthropocene as a time of crises' in this chapter), which is especially prominent in the information age and the emerging post-information age in which we live, I shall now introduce the potential of ecolinguistic research in studying and changing the dominant thinking related to sustainability.

Targeting Unsustainability Through Multimodal Ecolinguistic Research

Employing an empiricist approach, Halliday (2001, p. 195) viewed language as 'a theory of experience; a theory that is born of action, and therefore serves as a guide to action, as a metalanguage by which we live'. Emphasising the role of language with regard to how humans relate to the social and natural worlds, he further argued that 'classism, growthism, destruction of species, pollution [...] are not just problems for the biologists and physicists. They are problems for the applied linguistic community as well' (p. 199). Thus, in a Hallidayan understanding, language (which stands for social semiosis in the broader sense) functions beyond the role of information exchange; it plays a key role in how we conceptualise experience (i.e., the world perceived). Therefore, it influences the ways humans relate and respond to their experiences, that is, how they react to and act in the world.

Halliday's ideas go hand in hand with the prominent view in critical discourse studies that language plays a key role as a medium for constructing social and political realities in human societies. This view is based on understanding language/semiosis as action and language use as politics (see e.g., Fiske 1994; Chilton 2004; Chilton and Schäffner 2002; Fairclough 1989, 2003). The view that language/semiosis is positioned can be exemplified by how human interpretations of the world are based on frames (i.e., culturally embedded cognitive models on perceptions of reality) construed by the semiotic representations of *experience* (see Goffman 1974; Lakoff 2004; Tannen 1993; Tversky and Kahneman 1986). Frames, as shared elements of social cognition, promote particular ways of seeing an issue, and they appear in semiosis through a choice of linguistic features that represent, for example, a particular event, usually purposefully from a specific perspective. As an example, Shanahan (2007, pp. 2–3) listed the following climate change frames: 'scientific uncertainty frame, national security frame, polar bear frame, money frame, catastrophe frame, justice and security frame'. Fløttum and Gjerstad (2017) understood the connection between frames and narratives in terms of the roles of frames as storylines that communicate the reason that an issue represents a problem, what should be done, and the possible roles of the entities involved. At the same time as generating narratives, some frames also silence or background others and legitimise particular forms of social action (see van Dijk 2016). Therefore, frames motivate the actions that we may take and sometimes influence our inaction. They can also explain how we situate ourselves in the ecosphere. To locate underlying metanarratives, the identification and analysis of frames in semiotic structures help in establishing key value systems that are taken for granted or naturalised (see Barthes 1972; Fairclough 1995).

Thus, an ecolinguistic approach in the analysis of texts related to sustainability topics allows a critical exploration of the existing norms and values that guide the ways humans think and act in relation to the ecosystems they inhabit. Moreover, the study of other domains of meaning production beyond textual material—such as designs of living environments, landscapes and architecture—can expand such projects of critical exploration, as these domains represent material frames for human and non-human agency. As such, they epitomise the ways in which humans relate to the human and non-human components of their environment, for example, by indicating their preference for maximising production or human comfort instead of minimising the human footprint or avoiding harm to other-than-human beings. Nevertheless, studies extending the ecolinguistic approach to analysing material domains have been lacking, partly due to the shortage of methods encompassing such domains. This situation, however, has changed due to the growing availability of linguistic-semiotic methods equipped to study materials beyond (multimodal) textual data. This expansion is partly due to Halliday's (1978) concept of social semiotics and the broad understanding of semiotic resources (see O'Halloran and Fei 2014), which has enabled frameworks for the analysis of meanings in written or spoken texts as well as images, film, spaces, displayed art and architecture (see O'Toole 1994). Geared towards understanding semiotic choices across different resources of different modes, systemic functional multimodal discourse analysis (O'Halloran et al. 2019) broadens possible applications of multimodal discourse analysis, for example, to study events or embodied actions. Moreover, Halliday's view on social semiotics has inspired multimodal approaches to studying ideologies and identities, as exemplified by Machin and Mayr's (2012) multimodal critical discourse analytic approach.

Therefore, in response to the dominant sustainable development paradigm discussed in section 'The Anthropocene as a time of crises', I maintain that non-anthropocentric and more sustainable ways of framing the relationships between humans and other parts of nature could also be understood and voiced by linguistic research targeting differently positioned frames in alternative sustainability narratives. These narratives could be explored in various textual, visual and material discourses of, for example, regenerative communities (e.g., global ecovillage networks and the permaculture movement), sentiocentric animal rights communities or farmed animal sanctuaries. Such new understandings of the relationships between humans and other parts of nature are also needed to overcome humanity's current hypocognition (i.e., lack of conceptualisation and related vocabulary) with regards to the environment, which Lakoff (2010) attributed to the complexity of issues related to anything environmental in human societies.

Conclusion

This chapter aimed to draw attention to the semiotic factors behind the crises of the Anthropocene and criticise the role of our current dominant understanding of sustainability in their maintenance. To look beyond sustainable development, I pointed to alternative ways of seeing sustainability in terms of the ways in which humans relate to other parts of the environment, especially other sentient beings. Finally, I touched on a possible new area in ecolinguistics that may have the potential to reveal non-anthropocentric alternatives of the roles of humans in the natural world and expand our views on sustainability. In light of the scale of the crisis, however, I share Haraway's (2016b), Silova's (2021), Stibbe's (2024) and Ürge-Vorsatz's (2023) views on the subject: even with drastic changes in our approach and ways of living, massive losses of lives, species, habitats and livelihoods are inevitable. We have to learn to *stay with the trouble,* face the losses, mourn and make new connections with human and non-human others. We have to strive for change by the means available to each of us.

References

Althusser, L. 1971. Ideology and ideological state apparatuses. In *Lenin and philosophy and other essays*, trans. Ben Brewster. New York: Monthly Review Press.

Bar-On, Y.M., R. Phillips, and R. Milo. 2018. The biomass distribution on Earth. *Proceedings of the National Academy of Sciences* 115 (25) (June 19): 6506–6511. https://doi.org/10.1073/pnas.1711842115.

Barthes, R. 1972. *Mythologies*, trans. Annette Lavers. London: Jonathan Cape.

Boscardin, L., and L. Bossert. 2015. Sustainable development and non-human animals: Why anthropocentric concepts of sustainability are outdated and need to be extended. In *Ethics of science in the research for sustainable development*, ed. S. Meisch, J. Lundershausen, L. Bossert, and M. Rockoff. Baden-Baden: Nomos Publishing House.

Brundtland, G. 1987. *Our common future.* Oxford: Oxford University Press. Available from: https://sustainabledevelopment.un.org/content/documents/5987our-common-future.pdf (accessed August 30, 2024).

Caimotto, M.C. 2024. Discourses of cycling advocacy and power amidst wars, petro-masculinity and climate inaction. In *Ecological Communication and Ecoliteracy: Discourses of Awareness and Action for the Lifescape*. Ed. by M. Bortoluzzi, and E. Zurru. London: Bloomsbury advances in ecolinguistics.

Chau, M.H., C. Zhu, G.M. Jacobs, N.L. Delante, A. Asmi, S. Ng, S.S. John, Q. Guo, and K. Shunmugam. 2022. Ecolinguistics for and beyond the Sustainable Development Goals. *Journal of World Languages* 8 (2) (October 3): 323–345. https://doi.org/10.1515/jwl-2021-0027.

Chilton, P.A. 2004. *Analysing political discourse: Theory and practice.* Oxfordshire: Routledge.

Chilton, P.A., and C. Schäffner. 2002. Introduction: Themes and principles in the analysis of political discourse. In *Politics as text and talk: Analytic approaches to political discourse*, ed. P.A. Chilton, and C. Schäffner. Amsterdam: John Benjamins Publishing Company.

Crist, E. 2019. *Abundant earth: Toward an ecological civilization.* Chicago: University of Chicago Press.

Daggett, C. 2018. Petro-masculinity: Fossil fuels and authoritarian desire. *Millenium* 47 (1): 25–44. https://doi.org/10.1177/0305829818775817.

Du Plessis, C., and P. Brandon. 2015. An ecological worldview as basis for a regenerative sustainability paradigm for the built environment. *Journal of Cleaner Production* 109 (December): 53–61. https://doi.org/10.1016/j.jclepro.2014.09.098.

Fairclough, N. 1989. *Language and power*. London: Longman.

Fairclough, N. 1995. *Critical discourse analysis.* London: Longman.

Fairclough, N. 2003. *Analysing discourse: Textual analysis for social research.* Oxfordshire: Routledge.

Fiske, J. 1994. *Media matters: Everyday culture and political change.* Minneapolis: University of Minnesota Press.

Fløttum, K., and Ø. Gjerstad. 2017. Narratives in climate change discourse. *WIREs Climate Change* 8 (1) (January): e429. https://doi.org/10.1002/wcc.429.

Gibbons, L.V. 2020. Regenerative—The new sustainable? *Sustainability* 12 (13) (July 7): 5483. https://doi.org/10.3390/su12135483.

Goffman, E. 1974. *Frame Analysis: An essay on the organization of experience*. Cambridge: Harvard University Press.

Halliday, M.A.K. 1978. *Language as social semiotic: The social interpretation of language and meaning*. London: Edward Arnold.

Halliday, M.A.K. 2001. New ways of meaning: The challenge to applied linguistics. In *The Ecolinguistics reader: language, ecology and environment*. London: Continuum International Publishing Group.

Haraway, D.J. 2016a. Tentacular thinking: Anthropocene, Capitalocene, Chthulucene. *E-Flux Journal* 75 (September 2016). Available from: https://www.e-flux.com/journal/75/67125/tentacular-thinking-anthropocene-capitalocene-chthulucene/ (accessed September 1, 2024).

Haraway, D.J. 2016b. *Staying with the trouble: Making kin in the Chthulucene*. Durham: Duke University Press.

Hickel, J. 2015. Five reasons to think twice about the UN's Sustainable Development Goals. Available from: http://blogs.lse.ac.uk/southasia/2024/09/11/five-reasons-to-think-twice-about-the-uns-sustainable-development-goals/ (accessed September 10, 2024).

Jalava, M., T. Kaarlenkaski, O. Latva, E. Nikkilä, T. Räsänen and T. Syrjämaa. 2024. Introduction: Towards a history of animal industries in the Nordic countries. In *Animal industries: Nordic perspectives on the exploitation of animals since 1860*. Berlin: De Gruyter. https://doi.org/10.1515/9783110787337-001.

JYU.Wisdom community. 2021. Planetary well-being. *Humanities and Social Sciences Communications* 8 (1) (November 2): 258. https://doi.org/10.1057/s41599-021-00899-3.

Kanerva, J., and A. Krizsán. 2021. Discouraging climate action through implicit argumentation: An analysis of linguistic polyphony in the Summary for Policymakers by the Intergovernmental Panel on Climate Change. *Discourse & Communication* 15: 609–628. https://doi.org/10.1177/17504813211026512.

Lakoff, G. 2004. *Don't think of an elephant! Know your values and frame the debate: The essential guide for progressives*. White River Junction: Chelsea Green Publishing.

Lakoff, G. 2010. Why it matters how we frame the environment. *Environmental Communication* 4 (1): 70–81.

Lyotard, J.-F. 1984. *The postmodern condition: a report on knowledge*, trans. G. Bennington, and B. Massumi. Minneapolis: University of Minnesota Press.

Machin, D., and A. Mayr. 2012. *How to do critical discourse analysis: A multimodal introduction*. London: Sage.

Martusewicz, R.A. 2018. EcoJustice for teacher education policy and practice: The way of love. *Issues in Teacher Education* 27 (2): 17–35. Available from: https://www.itejournal.org/wp-content/pdfs-issues/summer-2018/05martusewicz.pdf (accessed October 18, 2024).

Nervino, E., J.O. Cheung, and J. Chen. 2024. Charting the path of sustainability discourse research: A systematic review of applied linguistic studies. *International Journal of Applied Linguistics* 34: 862–883. https://doi.org/10.1111/ijal.12537.

O'Halloran, K.L., and V.L. Fei. 2014. Systemic functional multimodal discourse analysis. In *Interactions, images and texts: A reader in multimodality*. Berlin: De Gruyter Mouton.

O'Halloran, K.L., S. Tan and P. Wignell. 2019. SFL and multimodal discourse analysis. In *the Cambridge handbook of systemic functional linguistics*. Cambridge: Cambridge University Press.

O'Toole, L.M. 1994. *The language of displayed art*. Leicester: Leicester University Press.

Plumwood, V. 2001. *Environmental culture: The ecological crisis of reason*. New York: Routledge.

Rappaport, R.A. 1999. *Ritual and religion in the making of humanity*. Cambridge: Cambridge University Press. https://doi.org/10.1017/CBO9780511814686.

Rappleye, J., and H. Komatsu. 2020. Towards (comparative) educational research for a finite future. *Comparative Education* 56 (2): 190–217. https://doi.org/10.1080/03050068.2020.1741197.

Ricoeur, P. 1984. *Time and narrative* (vol. 1), trans. K. McLaughlin, and D. Pellauer. Chicago: University of Chicago Press.

Ricoeur, P. 1985. *Time and narrative* (vol. 2), trans. K. McLaughlin, and D. Pellauer. Chicago: University of Chicago Press.

Shanahan, M. 2007. Talking about a revolution: Climate change and the media. *IIED Briefing Papers*, December 2007. Available from: https://www.iied.org/17029iied (accessed November 14, 2024).

Silova, I. 2021. Facing the Anthropocene: Comparative education as sympoiesis. *Comparative Education Review* 65 (4) (November 1): 587–616. https://doi.org/10.1086/716664.

Smithers, R. 2017 Vast animal-feed crops to satisfy meat needs are destroying the planet. *The Guardian*, October 5, 2017. Available from: https://www.theguardian.com/environment/2017/oct/05/vast-animal-feed-crops-meat-needs-destroying-planet (accessed September 23, 2024).

Steffensen, S.V., M. Döring, and S.J. Cowley. 2024. Ecolinguistics: living and languaging. In *Language as an ecological phenomenon: languaging and bioecologies in human-environment relationships*. London: Bloomsbury Publishing.

Stibbe, A. 2019. Education for sustainability and the stories we live by. In *Prioritizing sustainability education: A comprehensive approach*. Oxfordshire: Routledge.

Stibbe, A. 2021. *Ecolinguistics: Language, ecology and the stories we live by*, 2nd ed. Oxfordshire: Routledge.

Stibbe, A. 2024. *Econarrative: Ethics, ecology and the search for new narratives to live by*. London: Bloomsbury Publishing.

Syrjämaa, T., M. Jalava, T. Kaarlenkaski, O. Latva, and T. Räsänen. 2021. UNSUS—Tutkimushanke eläinteollisuuden ja kestämättömän kehityksen historiasta. *Historia nyt!* 4: 40–43.

Tannen, D., ed. 1993. *Framing in discourse*. Oxford: Oxford University Press.

Torpman, O., and H. Röcklinsberg. 2021. Reinterpreting the SDGs: Taking animals into direct consideration. *Sustainability* 13: 843. https://doi.org/10.3390/su13020843.

Tremblay, D., F. Fortier, J.F. Boucher, O. Riffon, and C. Villeneuve. 2020. Sustainable Development Goal interactions: An analysis based on the five pillars of the 2030 Agenda. *Sustainable Development* 28: 1584–1596. https://doi.org/10.1002/sd.2107.

Tversky, A., and D. Kahneman. 1986. Rational choice and the framing of decisions. *The Journal of Business* 59 (4): 251–278.

UNESCO. 2024a. UNESCO World Water Assessment Programme. *The United Nations world water development report 2024: Water for prosperity and peace*. UNESCO, Paris. Available from: https://unesdoc.unesco.org/ark:/48223/pf0000388952 (accessed September 11, 2024).

UNESCO. 2024b. *UNESCO raises global alarm on the rapid degradation of soils*. Press release. Available from: https://www.unesco.org/en/articles/unesco-raises-global-alarm-rapid-degradation-soils (accessed September 12, 2024).

United Nations. 2015. *The 2030 agenda and the Sustainable Development Goals: an opportunity for Latin America and the Caribbean*. United Nations. Available from: https://repositorio.cepal.org/server/api/core/bitstreams/6321b2b2-71c3-4c88-b411-32dc215dac3b/content (accessed September 12, 2024).

Ürge-Vorsatz, D. 2023. Ilyen hely lehet Magyarország harminc év múlva. [Hungary can be such a place in 30 years.] By Sándor Márton Német. Index.hu. Available from: https://index.hu/belfold/2023/08/22/sziget2023-klimavaltozas-eghajlatvaltozas-globalis-felmelegedes-kornyezetvedelem-urge-vorsatz-diana-sziget-fesztival/?token=8e4a2649a57f33006bf29ce2ffdc9b 6925.9.2024 (accessed September 23, 2024).

van Dijk, T.A. 2016. Critical discourse studies: A sociocognitive approach. In *Methods of Critical Discourse Studies*, 3rd ed. Thousand Oaks: SAGE Publications.

Virtanen, P.K., L. Siragusa, and H. Guttorm. 2020. Introduction: Toward more inclusive definitions of sustainability. *Current Opinion in Environmental Sustainability* 43: 77–82. https://doi.org/10.1016/j.cosust.2020.04.003.

Visseren-Hamakers, I.J. 2020. The 18th Sustainable Development Goal. *Earth System Governance* 3. Available from: https://doi.org/10.1016/j.esg.2020.100047 (accessed September 7, 2024).

Part II

Pedagogy and Practices

5

Future Visions and Pedagogical Practices in Pre-service Language Teachers' Reflections on Sustainable Development in Language Teaching in Finland and Namibia

Salla-Riikka Kuusalu, Meameno Aileen Shiweda, Minna Maijala, Minna Maunula, Minna Maunumäki, and Soila Merijärvi

Introduction

Worldviews, historical backgrounds, and immediate surroundings can affect how one perceives sustainability (Bonnett 2002) and, consequently, how one envisions it in the future. Differences in schooling systems and environmental education levels can impact the selection of pedagogical practices considered suitable for teaching about sustainability. Furthermore, the limited resources available in Global South countries, such as Namibia, can hinder the successful implementation

S.-R. Kuusalu (✉) · M. Maijala · S. Merijärvi
School of Languages and Translation Studies, University of Turku, Turku, Finland
e-mail: srkbos@utu.fi

S.-R. Kuusalu
Department of Biology, University of Turku, Turku, Finland

M. Maijala et al. (eds.), *Transformative Language Teaching for Sustainability*,
https://doi.org/10.1007/978-3-031-85493-4_5

of the Sustainable Development Goals (SDGs; Singh 2020) set by the Agenda 2030 Action Plan (United Nations 2015). Pre-service teachers can be considered *agents of change* (Bürgener and Barth 2018; Weinberg et al. 2020), who play a crucial role in promoting the achievement of the SDGs and implementing Education for Sustainable Development (ESD; UNESCO 2020), which was designed to advance personal and societal transformation to help in the work against environmental degradation and the adverse effects of climate change, and its implementation is enforced among the UN member nations.

Teacher education is an important stage in the life of becoming teachers, as it builds professional identity and reflects on one's own experiences and perspectives for the future (Steadman et al. 2018). In teacher education, the integration of future thinking and sustainable development plays a crucial role in preparing pre-service teachers to meet the challenges of the future (Kangas et al. 2024). Language teaching should provide opportunities to practise future-oriented thinking and help students envision alternative futures (see Varpanen et al. 2022). Therefore, pre-service language teachers should occasionally reflect on their own future (Henry 2020) and their own ability to prepare their students for future challenges, which can be very unpredictable. Furthermore, awareness of one's own competence in sustainability education helps subject teachers to implement it in their classes (Uitto and Saloranta 2017). Therefore, both the content and context of teacher training as well as the prestige of the teaching profession can affect the implementation of ESD and thus, the promotion of SDGs in the future work life of teachers.

M. A. Shiweda
Hifikepunye Pohamba Campus, University of Namibia, Windhoek, Namibia

M. Maunula
Kokkola University Consortium Chydenius, University of Jyväskylä, Jyväskylä, Finland

M. Maunumäki
Open University, University of Jyväskylä, Jyväskylä, Finland

Namibian and Finnish teacher training systems differ in multiple ways, and for instance, the status of teachers is higher in Finland (Tirri 2014) than in Namibia (Junias et al. 2022), resulting in differences in the backgrounds of teacher trainees. Additionally, there are structural differences in language teacher training. In Namibia, the integration of environmental sustainability into the curriculum is guided by the Constitution (Article 95) and the official national curriculum for basic education (MEFT 2019), and thus ESD is integrated into the curriculum and environmental education as a cross-curricular theme (UNESCO 2020). However, there are still quite a few challenges due to lack of awareness and lack of mainstreaming of multilingual practices in education (Clegg and Afitska 2011; Norro 2022), and teacher education could provide teacher trainees with an even more meaningful contribution to promote ESD in their work (Mpofu-Hamadziripi et al. 2022). That said, there have been some noticeable changes in the transformed University of Namibia curriculum, regarding the inclusion of ESD modules. Finland, on the other hand, has emphasised quality education and high-level skills as the key to achieving the SDGs (Ministry of Education and Culture 2022). Although the Finnish national curriculum for basic education was published in the previous decade, it already clearly paved the way to the strengthening of the role of sustainability education (Lähdemäki 2019; EDUFI 2014). Sustainability-related issues are highlighted both in the national curriculum for basic education and for upper secondary schools in Finland, particularly in the description of transversal or cross-curricular competencies and learning modules (EDUFI 2014, 2019).

The quantity of research on promoting sustainability in teacher education has significantly increased in the past decade. Specifically, there has been a focus on the inadequate practical skills of student teachers in training their own students to become responsible actors for the future (e.g., Dahl 2019; Pegalajar-Palomino et al. 2021). However, their visions of a sustainable future for language teaching in practice, or of realising a sustainable future in their work, have not been studied as far as we know. Studies have shown that teacher trainees are willing to include sustainability-related content in teaching but do not necessarily have skills on how to implement this (e.g., García-González et al.

2020; Maijala et al. 2023b). Training has also been shown to improve students' knowledge and awareness of sustainability (Sommier et al. 2022; García-González et al. 2020). This study offers a novel perspective and a comparative view on language education by exploring how pre-service language teachers envision sustainable development's role in future language teaching and what pedagogical practices they see as sustainable and viable for promoting ESD in two markedly different contexts.

Theoretical Background

Context Differences Between Finland and Namibia

In the Finnish context, students must complete a year-long teacher training programme through the departments of teacher education at university-affiliated teacher training schools in addition to their subject studies through corresponding university language departments (Jyrhämä et al. 2008). The university language departments provide subject-matter expertise, while the departments of teacher education and the teacher training schools provide pedagogical content expertise and practical skills (e.g., Pollari et al. 2018). Research-based teacher education can be regarded as one reason Finland has succeeded in the OECD Programme for International Student Assessment (Sahlberg 2018). The communicative language teaching approach, which involves learning languages through interaction, has been widely considered the primary methodological approach to language teaching for several decades (e.g., Ellis 2012). This approach forms the basis for the Common European Framework of Reference for Languages (Council of Europe 2001, 2020) and is used in the curricula of many European countries, including Finland (e.g., EDUFI 2014). The use of dialogues as content of language textbooks reflects the communicative approach (e.g., Maijala and Tammenga-Helmantel 2019).

In Finnish language classrooms, textbooks are widely used (Harjanne et al. 2017) and language teachers tend to choose topics from textbooks (e.g., Hahl 2020). Although Finnish language teaching mostly relies on textbooks, their use is not necessarily in the focus of teacher education programmes (Maijala 2020). The current national curriculum in Finland places a strong emphasis on language awareness and the integration of different languages in education (EDUFI 2014). However, there are challenges in the implementation in becoming more linguistically aware and in promoting the use of students' linguistic resources in Finnish classrooms (Alisaari et al. 2019; Repo 2023). Compared to other European countries, the status of teachers in Finland has been relatively high and they have been regarded as trusted professionals (Paronen and Lappi 2018; Pollari et al. 2018). However, in recent years, the popularity of the teaching profession has diminished. According to a recent survey by the Trade Union of Education in Finland (OAJ 2021), over half of the Finnish teachers had considered changing their careers. The most common reasons for this were a high amount of work, increased workload, and a low level of pay compared to other professions (Heikkinen et al. 2020). Aspiring subject teachers, including language students, apply to one-year teacher education programmes at departments of teacher education after acquiring subject-matter knowledge in their majoring language from corresponding university departments. Pre-service teachers learn practical teaching skills in real classrooms supervised by experienced mentoring teachers in Finnish teacher training schools associated with universities (e.g., Pollari et al. 2018).

Project-based learning was suggested as a practice to implement cross-curricular teaching, such as sustainability (Lähdemäki 2019). Accordingly, Finnish teacher education has focused on cross-curricular activities (e.g., Lavonen 2018). For the implementation of ESD, the contents of language textbooks are crucial, because that is from where language teachers tend to select the topics they teach (e.g., Richards 2014; Tomlinson 2012). During the last decade, the use of digital textbooks and other digital teaching materials has increased. In general, the Finnish education system aims to develop teachers' digital literacy to enable the digital transformation of education, with a specific focus on developing teachers' digital skills and competencies (OECD 2023). Moreover, Ministry of Education and Culture (2023) has set a vision for Finland to lead in the development and use of sustainable digitalisation in the educational sector by 2027.

In Namibia, students enrol in a Bachelor of Education honours programme, which consists of two components, pedagogical and practical. The practical aspect entails teaching methodology, which includes students partaking in school-based education for primary students or a teaching practicum for senior secondary student teachers. Training programmes focus on teaching different languages and include pedagogical and linguistic courses. The emphasis is consistently on Namibia's multilingual and multicultural environment (Litre et al. 2022), but the quality of training and resources may vary locally. English was adopted as official language and a medium of instruction (MoI) at the nation's independence in 1990, and it is a second language to most Namibian learners. In the same year, communicative methods, which are seen as successful in Western nations, were also given preference in Namibia to help learners for whom English is a second or third language improve their language proficiency (O'Sullivan 2001). However, a recent study on how educators deliver their classes discovered that educators were just utilising conventional techniques to teach English language (Ding et al. 2019). In language teaching, English, like any other language, is considered to possess the potential to act as a catalyst for personal growth and to assist in the development of positive attitudes, moral values, critical thinking abilities, broad general knowledge, and aesthetic sensibilities for learners (NIED 2016). English is mainly used in most classrooms, places where English is most prominent. However, English usage ends there as most pupils and students tend to communicate in their mother tongues. Although it is expected that teachers' English command should enable them to use English as an MoI, it is found that teachers were not effective in influencing the language abilities of their learners due to their own shortcomings such as the lack of teaching skills and content knowledge (Ngololo and Nekongo-Nielsen, 2017; Norro 2021).

In most Namibian classes, a textbook is regarded as *an indispensable resource* as many teachers are considered unable to teach without it (Republic of Namibia 2008). The usage of textbooks is supported by The Namibian National Textbook Policy (MAWF 2008), which states that textbooks can significantly affect learners' academic performance, especially when effectively utilised by a qualified or experienced teacher. In some instances, however, teachers improvise and use different sources when teaching English as a second language (ESL), use learner centred teaching methods and teamwork, and make communication in ESL compulsory for the learners to help them develop their vocabulary (Haufiku 2022). Developing a student's comprehension of knowledge and skill acquisition through ICT-based learning helps to lay the groundwork for a society built on knowledge (Ministry of Education 2010), which can lead to participatory teaching and learning methods that motivate and empower learners to change their behaviour and act for sustainable development (Shava et al. 2021). Within Namibian educational contexts, digital literacy is apparent, owing partly to the efforts of the governmental Digital Namibia Strategy, which has aimed to strengthen areas of digital transformation since 2021 (Mare et al. 2023; Mbangula 2024). Furthermore, other stakeholders, like the UN Resident Coordinator in Namibia (UN 2022), hope that their efforts to use ICTs in education will help achieve Goal 17 of Agenda 2063's aspiration 6, which aims to remove all obstacles to high-quality social, health, and educational services across the African continent.

Research on Pre-service Language Teachers' Views on Implementing Sustainability

Research on the perceptions of aspiring language teachers on sustainability and its education is still scarce, but some studies have been conducted recently in the Nordic context. Maijala et al. (2023b) discovered that language teacher trainees in Finland considered social and cultural aspects of sustainability to be the most relevant in language teaching. In an earlier study, Laine et al. (2022) studied Finnish preservice language teachers' perceptions on teaching about sustainability using a questionnaire and a follow-up group reflection and found out that the students valued including sustainability in language classes but felt uncertain about their qualifications, lack of planning time, and subject knowledge. In a study by Kuusalu and others (2024), students tended to approach sustainability themes by prioritising global rather than local issues but could change their approach quite randomly between the sustainability dimensions. This is consistent with what

Spiropoulou and others (2007) found early on in their study on primary teachers' attitudes on ESD: local and national issues are highlighted while global issues get easily neglected. Further, the practices language teacher students suggested for teaching about sustainability have been mainly teacher-centred (Maijala et al. 2023b), and thus, not ideal for the incorporation of ESD. Based on the conclusions of these earlier studies, from the point of view of ESD, language teacher education in the Nordics would need to develop towards better supporting teachers' competencies in systems thinking, student-oriented pedagogy, and overall sustainability knowledge.

Research on Future Visions of Pre-Service Teachers

Some research has been conducted in the Finnish context on pre-service language teachers' future visions, particularly regarding their future work as teachers, but research in Namibian or related contexts is still lacking. Ruohotie-Lyhty and Pitkänen-Huhta (2022) conducted a multimodal study on the perspectives of students on their future as language teachers. They identified both optimistic and pessimistic views of the future profession, but descriptions of the dream job scenario often included a method supportive of sustainability education, where the teacher would sit among the students and converse with them. In Maijala's (2023a) study, visual and written data were collected from language teacher students at the beginning and end of their pedagogical studies. The students wrote and drew a postcard of their future, which was reflected on at the end of the studies. The results revealed that some pre-service teachers experienced uncertainty about the future due to insecurity of employment prospects and their own digital skills. The focus on the development of teachers' digital literacy skills was demonstrated by Maijala et al. (2023a), who examined pre-service language teachers' perceptions of future learning environments. The participants were asked to send an email from the imaginary future in which they depicted their daily work as a language teacher in 2040. The findings revealed that pre-service language teachers associated learning environments with new teaching methods offered by new technologies and the flexibility, interdisciplinary, and collaboration these allow. They viewed technology as a means of achieving better inclusion through distance learning and participatory methods.

There are still many challenges in developing student teachers' thinking about the future in Finland. This is related to the fact that most Finnish teacher education programmes are built around subject-specific didactics studies, and social and contextual issues are considered less important (Toom and Husu 2021; Juvonen and Toom 2023). According to Säntti et al. (2023), decontextualisation is manifested in Finnish teacher education as the decline in so-called contextualised studies, such as the history of education, as the focus of teaching and textbooks is at the classroom level. This may be a factor why teachers often remain adaptors to the prevailing situation rather than become active agents of social change.

Materials and Methods

We used the following questions to direct the research during the course of this study:

1. What aspects of sustainable development do pre-service language teachers consider relevant in their future visions of language teaching?
2. What kind of pedagogical practices do pre-service language teachers consider beneficial in promoting sustainable development in language teaching?

Participants

The study involved Finnish ($N = 39$) and Namibian ($N = 33$) pre-service language teachers with seven Namibian class teachers, who teach language subjects. The participants' ages ranged from 22 to 32 years with a gender distribution of 33 women and 6 men in Finland. The Namibian participants' ages ranged from 22 to 36 years with 21 being women and

12 men. All 39 Finnish students reported Finnish as one of their native languages, with 36 of them considering it their sole native language. The other mentioned home languages were German, Russian, and Swedish. In contrast, the Namibian participants reported five home languages: Oshindonga ($n = 14$), Oshiwambo ($n = 10$), Oshikwanyama ($n = 7$), Otjiherero ($n = 1$), and Afrikaans ($n = 1$), but only one person stated several home languages. The participants hailed from various regions in northern Namibia, including Oshana, Omusati, Oshikoto, and Ohangwena. Most of the Finnish participants came from different regions across Finland, Kajaani being the northernmost area. Two learners in the Finnish group originated from countries other than Finland.

During the time of the survey, the Finnish pre-service teachers were at the end stage of their studies and 11 of them had already completed their teacher training, while 27 were currently undergoing it. Additionally, one learner from the Finnish group had completed basic studies (25 ECTS) in sustainable development as a minor subject, and two others had taken individual courses related to sustainability. The Namibian participants, on the other hand, were in their final (4th) year of becoming subject teachers. They taught English and one of the two official Indigenous languages (Oshikwanyama or Oshindonga), as well as social sciences. Notably, ten Namibian participants had completed basic studies in sustainable development, and an additional four students had taken individual courses focused on sustainability.

Data Collection

Researchers employed a questionnaire to collect data on the future visions and pedagogical practices of Finnish and Namibian pre-service language teachers. This same questionnaire is part of a broader research effort that also investigates how pre-service students perceive sustainable development aspects beyond the teaching context (Kuusalu et al. 2025). The Webropol-based questionnaire consisted of basic inquiries about participants' backgrounds (age, gender, home tongue[s]), along with an essay assignment. Participants were asked to write a brief essay on the significance of sustainable development in language learning. The instructions included specific questions related to sustainability, both in general and within the context of language teaching. Additionally, participants were prompted to consider their own future prospects as pre-service teachers and how SDGs could be promoted in language teaching. This study specifically focuses on the responses related to the following directional questions within the questionnaire:

- How can language teachers promote Sustainable Development Goals?
- How will issues related to sustainable development become evident in language teacher's work in the future?

Data from both Namibian and Finnish pre-service teachers were collected with the questionnaire during an academic seminar meeting. This approach ensured that learners were more likely to respond when introduced to the questionnaire during the school day. Participants had the option to allow or deny the use of their responses for research purposes. They were informed about the anonymity clause, and then given up to 60 minutes to complete the questionnaire. In Namibia, some participants opted for a printed version of the questionnaire due to internet connectivity issues. All essays underwent scrutiny to detect any signs of search engine or AI usage, and certain responses were excluded from the dataset due to plagiarism. This issue was particularly pronounced in the essay answers provided by Namibian learners. Finnish learners, however, did not directly copy text into their responses, although the possibility of internet searches could not be completely ruled out. Responses were collected anonymously using identification numbers, and the data was securely stored in the university's cloud storage. During data analysis, the ID numbers were used to prevent participant recognition based on other aspects of the data, such as response time. Finnish learner responses were classified as ID numbers 1–39, while Namibian learners were assigned ID numbers 40–72.

Data Analysis

The responses were analysed using a summative content analysis approach (Hsieh and Shannon 2005). The analysis focused on what future visions the reflections of the Namibian and Finnish pre-service teachers entailed, and what type of pedagogical practices were mentioned. The responses to the former question emphasised future either by stating the phrase *in the future* or with the use of future tenses in English or in Finnish; the responses to the latter question mostly entailed the word *promote*, or otherwise expressed how language teachers could improve the promotion of ESD in language teaching. A comprehensive overview of the data prompted eight future vision categories and nine categories of different approaches to teaching ESD in the language learning context (see Table 5.1 for a breakdown of the categories).

The research team followed good scientific practice and ethics following the guidelines of Finnish national board on research integrity (TENK 2023) throughout the research process. An important ethical principle was to value the environment, ecosystems, cultural heritage, and societies. This was important because the research team combined the perspectives of the Global North and South, but the majority of researchers were from the Global North. Therefore, analyses and interpretations were discussed and clarified in researcher triangulation meetings (Patton 2015). The research process aimed to show respect for fellow researchers, their cultures and societies, and ensure the validity of the research.

Results

In this section, first the aspects and categories of sustainable development that pre-service language teachers consider relevant in their future visions of language teaching are presented. Next, the categories of pedagogical practices that pre-service language teachers consider to be beneficial in promoting sustainable development in language teaching are shown. The results include both qualitative quotes and quantitative percentages from both the Finnish and Namibian student teacher data.

Table 5.1 Category definitions

Future visions	Pedagogical practices
Considering equality Equality issues are addressed as a future concern.	*Action-oriented teaching* Entails action-oriented teaching approaches and visions, such as immersion activities.
Digitalisation Teaching materials will become digitised even more, and/or the use of digital devices will increase.	*Cross-curricularity and theme days* Cross-curricularity is implemented, which means combining two or more subjects into one in a lesson plan, and incorporating theme days into language teaching.
Laws and practices Laws and practices concerning schools, administration, and/or teaching will be altered.	*Cultural knowledge* Teaching about sustainable development is done through the cultures of the target language and focusing on the cultural dimension.
Linguistic sustainability Languages are preserved, and their futures are considered on the spoken and written level.	*Implicit learning* Teaching about sustainable development is done implicitly without making the theme too obvious to the learners, meanwhile implementing it as pieces to lesson plans.
Grass-root level Sustainability will be visible at the grass-root level, meaning through teachers, curricula, and teaching materials.	*Leading the change* Teacher focuses on their own actions from the perspective of sustainable development and understands how their own actions are perceived by learners and encourages learners to act.
Student engagement Learners will and should change from passive to active agents in discussion and actions.	*Lifelong learning* Teacher focuses on further training and keeps themself updated about new pedagogical approaches.
Teacher education Issues regarding sustainable development will be more visible in teacher education.	*Pedagogical materials* The importance of pedagogical materials is highlighted in the promotion of sustainable development, such as the use of digital devices.

(continued)

Table 5.1 (continued)

Future visions	Pedagogical practices
Increase in visibility Sustainability will become more visible in the future of language teaching and learning.	*Target language* The subject is approached through target language, such as vocabulary and speaking-skills exercises.
[No answer] No consideration of the future aspect was given.	*Enabling societal meaning and impact* Teachers enable learners to link learning outcomes to the surrounding society/world.

Future Visions

Finnish and Namibian language teachers considered the broad aspects of sustainable development that were important for language teachers in their future vision of language teaching. Table 5.2 shows the categorised mentions of the topic and the quantitative incidence.

The highest number of responses in future visions were *increase in visibility* (see Table 5.2), which was mentioned in the responses of Finnish (44%) and Namibian (24%) respondents. They expressed that: 'Issues related to sustainable development will become even more apparent in the future as awareness grows, and global efforts continue to address

Table 5.2 Percentages and number of mentions of future visions categories in the responses of Finnish and Namibian pre-service language teachers

Category	Total of mentions	Finland ($n = 39$)	Namibia ($n = 33$)
Increase in visibility	25	44% (17)	24% (8)
Linguistic sustainability	11	8% (3)	24% (8)
Teacher education	11	13% (5)	18% (6)
Grass-root level	10	13% (5)	15% (5)
Considering equality	7	8% (3)	12% (4)
Digitalisation	6	15% (6)	0% (0)
Student engagement	5	5% (2)	9% (3)
Laws and practices	2	3% (1)	3% (1)
No category mention	13	7	6

them' (ID64). According to the students, the increased visibility of sustainability would also be reflected in the work of language teachers in the future. Finns (8%) and Namibians (24%) also considered the aspect of *linguistic sustainability* by raising concerns over the strengthened position of dominant languages and the danger of disappearance of smaller languages and cultures. However, according to them, 'there is now a serious attitude towards the endangered status of other languages' (ID28). They reflected on how different language learning choices were linked to future sustainability: 'What language do we teach our students and why? Are the languages we choose sustainable and ecological? What are the "languages of the future"?' (ID35).

According to Finnish (13%) and Namibian (18%) pre-service teachers, teachers' role can be strengthened in the future by improving *teacher education*, particularly by integrating sustainability perspectives. They were concerned about the lack of sustainability skills in the education they received (see Example 1):

(1) I would like to see more emphasis on these issues in teacher training in the future, because otherwise they are left almost completely open, because the term itself is so abstract that it is difficult to even understand what should be done in practice. (ID11)

They also expressed their hopes and expectations for future teacher education, reflecting the socially relevant role of teachers (see Example 2).

(2) Language teachers training requires a lot of effort and dedication to develop and build necessary skills and knowledge in order to succeed and develop yourselves and the community as we grow in our learning experience plays a major role as well as in our future as future generations we carry a huge responsibility for learning, developing and improving our future. (ID45)

Both Finns (13%) and Namibians (15%) stressed the importance of the *grass-root level* aspect for future sustainable development. Teachers said sustainability will become 'even more important part of the curriculum and included in more teaching as opposed to individual English courses they are taught on now' (ID03). In the future, the work of language teachers will essentially involve aspects of sustainability, and the students felt it was appropriate to include this in the curriculum and also to anticipate the future.

The category *considering equality* was addressed as a future concern by Finnish (8%) and Namibian students (12%). According to the perspective of some Namibian students, individuals' awareness of sustainability would strengthen the stewardship of natural resources, which would have a global impact on the sustainability of the planet (see Example 3).

> (3) In future this [sic] issues can be seen significant as it will eradicate poverty if people do look after their natural resources, it will promote employment and publicity of our country and manage the earth's resource base for the benefit of all. (ID53)

In addition to the perspective of global equality, it was suggested that in the future it was also important to recognise students' different backgrounds as teachers. The starting points for students were identified as 'differences in backgrounds, e.g., language skills, socio-economic background, religion, etc., as well as in school skills' (ID22). However, a teacher could 'enable different learners to have similar rights to learning in a way that is as sustainable as possible in all aspects of sustainable development—economically, ecologically, socially and culturally' (ID22).

Finnish pre-service teachers mentioned digitalisation as one aspect of future sustainability in six responses. According to them 'we already use a lot of electronic platforms to teach and learn' (ID27) and 'for language teachers, textbooks will possibly go completely electronic and nothing will be done on paper' (ID16). Further, Finnish teachers reflected that digital materials would increase in the future in the teaching of all subjects and 'there will be new ways to incorporate them in the learning process' (ID27). Namibians ($n = 33$) did not mention digitalisation in their responses.

Pre-service teachers also mentioned *student engagement* as a relevant aspect of their vision for sustainable future language teaching. Both Finns (5%) and Namibians (9%) stressed that learners will and should change from passive to active agents in discussion and actions. They reflected: 'the idea of engaging students in a discussion on sustainability becomes significant' (ID04), and further 'by making sure our learners know about these issues through teaching and they practise sustainable development' (ID63). Pre-service teachers suggested that people in general are more aware of sustainability issues and the global perspective. One pre-service teacher further reflected on the objective from a language teaching perspective 'to form their own opinions, avoid misinformation and get active, they need to have the language for it' (ID31).

Namibian pre-service teachers were concerned about the decline in linguistic competence and the withering away of linguistic richness (see Example 4).

(4) To keep our language for the future generation not to be using foreign language only but their mother languages too. (ID67)

A few pre-service teachers, Finnish (3%) and Namibian (3%), also reflected on *laws and practices* as a reflection of the sustainability of future language teaching. This aspect was seen as external and binding on individuals 'there will be more laws and practices that we'll have to get used to' (ID09). The aspect of language teaching was emphasised as a way to improve the language learning environment in the future of teaching and learning, which would contribute to 'improved to provide the best results and if language needs of the people are sustained to a point whereby the advancement in sustainability can be seen' (ID55). A few students did not mention any aspect of the future sustainability of language teaching in their responses, among Finns ($n = 7$) and Namibians ($n = 6$).

Pedagogical Practices

The highest number of responses in pedagogical practices were *leading the change* ($n = 39$), *enabling societal meaning and impact* ($n = 34$), and *pedagogical materials* ($n = 22$; see Table 5.3). The category that was mentioned by both Finnish (44%) and Namibian (66%) pre-service teachers the most was teachers' role in *leading the change*. They emphasised that language teachers have the potential to contribute to sustainable development, as one pre-service teacher stated: 'Teachers can model sustainable practices in the classroom such as recycling, reducing paper usage' (ID59). In general, the sentences in this category often began with 'language teacher can/may/need to for instance try to encourage students' (ID08), 'improve students' awareness on sustainable development' (ID15) or 'to be aware of environmental issues' (ID20).

In Finnish pre-service teachers' responses, *cultural knowledge* belonging to the cultural dimension of sustainable development was highlighted (26%), whereas none of the Namibian responses linked this dimension to sustainable development when considering language teaching. Finnish students connected sustainable development related content with culture teaching, for instance, by using the target language and by telling learners about the target culture. Cultural diversity was also connected to the environmental dimension (see Example 5).

(5) Teaching about sustainable development also improves cultural awareness. Language teachers can play a key role in promoting sustainable development by helping their students to understand and appreciate the cultural and environmental diversity of the world. (ID15)

Table 5.3 Percentages and number of mentions of different pedagogical approaches in the responses of Finnish and Namibian pe-service language teachers

Category	Total of mentions	Percentage and number of mentions in Finland (n = 39)	Percentage and number of mentions in Namibia (n = 33)
Leading the change	39	44% (17)	66% (22)
Enabling societal meaning and impact	34	36% (14)	60% (20)
Pedagogical materials	22	44% (17)	15% (5)
Cross-curricularity and theme days	20	36% (14)	18% (6)
Target language	18	33% (13)	15% (5)
Cultural knowledge	10	26% (10)	0% (0)
Lifelong learning	9	3% (1)	24% (8)
Action-oriented teaching	8	15% (6)	6% (2)
Implicit learning	4	10% (4)	0% (0)
No category mention	1	0	1

In Namibian pre-service teachers' responses, the category *enabling societal meaning and impact* was underlined (60%), that is, teachers' role in enabling their students to link learning outcomes with the surrounding society or the world. The Namibian respondents emphasised the teacher's role and responsibility in educating future generations to protect and value their home environment.

Both Finnish and Namibian teacher trainees referred to the significance of community (see Example 6) and highlighted the importance of *cross-curricularity and theme days* and collaboration in the implementation of ESD (see Example 7).

(6) Teachers can also encourage students to take action in their own communities by participating in sustainable initiatives and promoting environmental awareness. (ID08)
(7) Cross-curricular modules on the subject [sustainable development] could also be implemented, allowing language teachers to collaborate with teachers from other subjects. (ID29)

The mentions in the category *pedagogical materials* stressed the importance of teaching materials in the implementation of ESD. Especially the Finnish pre-service teachers (44%) mentioned textbooks. Finnish teacher trainees (15%) also mentioned *action-oriented teaching* as a good method to implement topics of sustainable development into language classes. Their responses included action-oriented methods such as discussions, reflections, quizzes, and pair work.

In addition, students in both contexts mentioned that 'changing teaching methods and approaches during lesson preparation' (ID67) would make the topics more interesting and meaningful for the learners, as '[t]he whole idea of such activities is to make the students engage in ethical and responsible actions in terms of future prospects, enabling them to work on the macro scale' (ID04).

Target language category emerged in 33% Finnish and 15% Namibian responses. Mostly the target language was considered to be the medium with which some sustainability content should be discussed or otherwise learned in the language class, although some concern was raised because 'the topic has been talked about so much at this point that a lot of people have become bored of talking about it' (ID09).

Lifelong learning category came up especially in the responses of Namibian students (24%) when they offered suggestions on how to promote ESD in language classes, while only one Finnish student considered the issue by mentioning voluntary teacher training. In contrast, *implicit learning* was mentioned only by the Finnish pre-service teachers, although their number was small (10%). According to their responses, it would be important to have sustainable ideas and practices submerged into the language lessons, such that the learners could adopt the principles of sustainability without much effort.

Discussion

Respondents in both contexts emphasised how sustainability will become more prominent in language teaching and learning, which echoes the growing awareness of its importance in all areas of life (see Bonnett 2002). On the other hand, some of the respondents did not associate their future profession as language teachers with the sustainable future of language teaching. This can be interpreted such that the topic feels new to these pre-service language teachers and seemingly not related to their future profession. However, there is an increasing need for knowledge and competencies for sustainable development across all disciplines.

The reflections of pre-service teachers concerning linguistic sustainability partly reflected the cultural context, as in the Namibian context, English is the official language, but multiple other languages exist side by side. The Namibian participants in this study had altogether five home languages plus English as the language of instruction. All Finnish participants, however, were Finnish speakers, and even though linguistic variability has increased in Finland, the role of Finnish remains strong as the MoI. Therefore, it was quite expected that linguistic sustainability, such as concern over the continued existence of local languages and the use of home languages to assist learning at school (Norro 2022), arose primarily from the future visions of Namibian participants. The few Finnish students who showed awareness of these issues referred to the growing usage of English and the status of Sámi languages in Finland.

Interestingly, the students in both contexts identified the need for improving teacher education to integrate sustainability perspectives, expressed concerns about the lack of sustainability skills in current teacher training programmes, and emphasised the responsibility of teacher education for future generations (also Kangas et al. 2024). To answer these concerns, language teacher education might need to consider including social and contextual problematics in a more holistic, interdisciplinary manner (Toom and Husu 2021; Juvonen and Toom 2023). Work towards increasing the overall prestige of the teacher profession would likely encourage a wider student base and strengthen the capability of pre-service teachers to build their professional identity and practise their reflection skills (Steadman et al. 2018).

Quite many Finnish pre-service teachers considered digital language teaching tools and skills in their future visions; in Namibian pre-service teachers' responses, on the other hand, digitalisation was completely absent. This might be due to Namibian digital transformation being in an earlier phase than the Finnish one (Mbangula 2024), causing some unequal resource division between these educational settings (see Singh 2020). However, in both countries, the educational system strives for increasing digital literacy, as it is thought to support all learning and give learners more skills. Finnish pre-service teachers' viewpoints to sustainability were that of a responsible consumer of goods, whereas Namibian participants' views were more concretely linked to environmental issues, the surrounding nature, and natural resources. Both Finnish and Namibian respondents highlighted the importance of sustainability through teachers, curricula, and teaching materials. These results emphasising practicality are in line with previous studies (e.g., Uitto and Saloranta 2017; Pollari et al. 2018; Lavonen 2018; Lähdemäki 2019). Finnish respondents emphasised the significance of cross-curricular competencies in implementing ESD, probably because they are at focus in the Finnish national curricula (Lähdemäki 2019) and teacher education (Lavonen 2018). Positively, action-oriented teaching, a very supportive method for the goals of ESD, was acknowledged by pre-service language students. Both Finnish and Namibian teacher students recognised the dominant role of textbooks in promoting ESD. This reflects the fact that textbooks play a central role in Finnish

(Harjanne et al. 2017) and Namibian (Republic of Namibia 2008) language teaching and that pre-service language teachers use them extensively during their teacher training (Maijala 2020). In Chapter 6 of this volume, Maijala and others discuss their findings in a study of German language textbooks in several European contexts and highlight the benefits of incorporating the SDGs in textbooks and using language textbooks to integrate ESD in language classes.

Our study reveals a shared concern among pre-service teachers from Finland and Namibia regarding the integration of sustainability into future language teaching practices. Namibian pre-service teachers' responses showed concern about teachers' exemplary role in leading the change (Bürgener and Barth 2018; Weinberg et al. 2020) towards a more sustainable society and they saw it as a teacher's responsibility to enable their students to link learning outcomes with the surrounding society and the world. Finnish pre-service teachers elaborated on practical ways teachers can lead by example: using less paper, recycling materials, and biking to work. They preferred addressing local and tangible matters that are easily approachable. Similar results were obtained in earlier studies (Spiropoulou et al. 2007; Kuusalu et al. 2024), which suggests that language teachers may need more training to be able to confidently address global environmental problems in class. One of the greatest differences between the two contexts was that, unlike the Namibian respondents, Finnish students included cultural knowledge in their reflections as a means to teach sustainability through different cultures, for example by incorporating themes of cultural sustainability into language lessons. On the other hand, Namibian pre-service teachers elaborated on the importance of continuous education and training of teachers to keep up with the changing occupational requirements, an aspect mostly ignored by the Finnish participants.

The difference between the two contexts might be explained by the Finnish students' increased awareness of cultural sustainability in language teaching due to its recent inclusion in language teacher training in Finland (Laine et al. 2022). However, the content Namibian language teacher students learn does not entail cultural knowledge of the target language, hence there were no responses regarding this category. This lack of awareness may stem from language pedagogical knowledge that does not touch on cultural awareness about English as a language and could be addressed by adding aspects of cultural sustainability into Namibian language teacher training. More specifically, multilingual classroom practices and adaptable materials could be developed to increase the knowledge of cultural sustainability of language teachers in Namibia. This would significantly advance the abilities of language teachers in middle- and low-income postcolonial contexts to prepare their students to participate in the local decision-making processes.

Conclusions

There is a growing need to develop teacher education towards being more sustainable by, e.g., using blended learning opportunities together with transformative language teaching approaches to increase the knowledge base and sustainability competencies of future language teachers (Kangas et al. 2024). Based on our results, one way to address ESD in language teaching both in the Global North and the Global South would be to link sustainability themes more widely into language textbooks and other instructional materials considering the local contexts. In teacher training, sustainability could be supported by providing more guidance to teacher trainees on how to integrate sustainable development content and transformative teaching methods into their lessons and further, providing them with skills to enable their own change agency. A short training course on sustainability cannot give pre-service teachers all the sustainability skills a language teacher should have. Instead, principles of sustainability should be incorporated into the very core of teacher training programmes. Language teacher training is an essential starting point in the development of teachers' professional growth, attitudes, and values, whose effects reach far into working life (Hays and Reinders 2020; Kangas et al. 2024). Language teachers are valuable in the wide network of actors striving towards a global sustainability transformation, and language teacher education should acknowledge their potential. Unequal distribution of global resources can hinder local progress towards sustainability, which is why continued, multidisciplinary, and collaborative research among different contexts remains crucial for a mutual understanding of how to change societies for the better.

References

Alisaari, J., L.M. Heikkola, N. Commins, and E.O. Acquah. 2019. Monolingual ideologies confronting multilingual realities. Finnish teachers' beliefs about linguistic diversity. *Teaching and Teacher Education* 80 (April): 48–58. https://doi.org/10.1016/j.tate.2019.01.003.

Bonnett, M. 2002. Education for sustainability as a frame of mind. *Environmental Education Research* 8, no. 1 (February): 9–20. https://doi.org/10.1080/13504620120109619.

Bürgener, L., and M. Barth. 2018. Sustainability competencies in teacher education: Making teacher education count in everyday school practice. *Journal of Cleaner Production* 174 (February): 821–826. https://doi.org/10.1016/j.jclepro.2017.10.263.

Clegg, J., and O. Afitska. 2011. Teaching and learning in two languages in African classrooms. *Comparative Education* 47, no. 1 (February): 61–77. https://doi.org/10.1080/03050068.2011.541677.

Council of Europe. 2001. *Common European framework of reference for languages: Learning, teaching, assessment*. Strasbourg: Council of Europe. Available from: https://rm.coe.int/1680459f97. Accessed June 6, 2024.

Council of Europe. 2020. *Common European framework of reference for languages: Learning, teaching, assessment–Companion volume*. Council of Europe Publishing, Strasbourg. www.coe.int/lang-cefr.

Dahl, T. 2019. Prepared to teach for sustainable development? Student teachers' beliefs in their ability to teach for sustainable development. *Sustainability* 11, no. 7 (April 3): 1993. https://doi.org/10.3390/su11071993.

Ding, A.-C. E., A. Ottenbreit-Leftwich, Y.-H. Lu, and K. Glazewski. 2019. EFL teachers' pedagogical beliefs and practices with regard to using technology. *Journal of Digital Learning in Teacher Education* 35, no. 1 (January 2): 20–39. https://doi.org/10.1080/21532974.2018.1537816.

EDUFI (Finnish National Agency for Education). 2014. *Perusopetuksen opetussuunnitelman perusteet 2014* [Core curriculum for basic education 2014]. Helsinki: Opetushallitus.

EDUFI (Finnish National Agency for Education). 2019. *Lukion opetussuunnitelman perusteet 2019* [Core curriculum for general upper secondary education 2019]. Helsinki: Opetushallitus.

Ellis, R. 2012. *Language teaching research and pedagogy*. Chichester: Wiley-Blackwell.

García-González, E., R. Jiménez-Fontana, and P. Azcárate. 2020. Education for Sustainability and the Sustainable Development Goals: Pre-service teachers' perceptions and knowledge. *Sustainability* 12, no. 18 (September 18): 7741. https://doi.org/10.3390/su12187741.

Hahl, K. 2020. Kulttuurin muuttunut olemus opetussuunnitelmissa ja vieraiden kielten oppikirjoissa [The changed nature of culture in curricula and language textbooks]. In *Kielididaktiikan katse tulevaisuuteen: Haasteita, mahdollisuuksia ja uusia avauksia kielten opetukseen*, ed. R. Hilden, and K. Hahl, 173–201. Helsinki: University of Helsinki.

Harjanne, P., C.D. Larenas, and S. Tella. 2017. Foreign-language teaching and studying in Chilean and Finnish classrooms as seen by teachers. *Journal of Language and Cultural Education* 5, no. 3 (September 1): 1–21. https://doi.org/10.1515/jolace-2017-0025.

Haufiku, I. 2022. *Teaching challenges and coping strategies of English second language teachers in senior secondary schools in the Ohangwena region, Namibia*. Doctoral dissertation, University of Namibia.

Hays, J., and H. Reinders. 2020. Sustainable learning and education: A curriculum for the future. *International Review of Education* 66, no. 1 (February): 29–52. https://doi.org/10.1007/s11159-020-09820-7.

Heikkinen, H.L., J. Utriainen, I. Markkanen, I., Pennanen, M. Taajamo, and P. Tynjälä. 2020. *Opettajankoulutuksen vetovoima*. Available from: http://urn.fi/URN:ISBN:978-952-263-818-2. Accessed April 13, 2024.

Henry, A. 2020. Possible selves and personal goals: What can we learn from episodic future thinking? *Eurasian Journal of Applied Linguistics* (December 1): 479–498. https://doi.org/10.32601/ejal.834659.

Hsieh, H.-F., and S.E. Shannon. 2005. Three approaches to qualitative content analysis. *Qualitative Health Research* 15, no. 9 (November): 1277–1288. https://doi.org/10.1177/1049732305276687.

Junias, K., L. Kambeyo, and K.M. Katukula. 2022. An analysis of a teacher education programme at a Namibian tertiary institution. *Creative Education* 13 (8): 2635–2657. https://doi.org/10.4236/ce.2022.138168.

Juvonen, S., and A. Toom. 2023. Teachers' expectations and expectations of teachers: Understanding teachers' societal role. In *Finland's famous education system: Unvarnished insights into Finnish schooling*, ed. Thrupp M.,

P. Seppänen., J. Kauko, and S. Kosunen, 121–135. Singapore: Springer. https://doi.org/10.1007/978-981-19-8241-5.

Jyrhämä, R., H. Kynäslahti, L. Krokfors, R. Byman, K. Maaranen, A. Toom, and P. Kansanen. 2008. The appreciation and realisation of research-based teacher education: Finnish students' experiences of teacher education. *European Journal of Teacher Education* 31, no. 1 (February): 1–16. https://doi.org/10.1080/02619760701844993.

Kangas, J., M. Maunumäki, M. Maunula, and H. Harju-Luukkainen. 2024. Critical consideration of sustainable teacher education in Finland. *Pedagogy in the North: shifting concepts, altered states and common expressions* 2, no. 31. https://doi.org/10.26203/K107-3324.

Kuusalu, S.-R., P. Laine, M. Maijala, M. Mutta, and M. Patzelt. 2024. University language students' evaluations of ecological, social, cultural and economic sustainability and their importance in language teaching. *International Journal of Sustainability in Higher Education* 25, no. 9 (January 1): 1–18. https://doi.org/10.1108/IJSHE-05-2023-0169.

Kuusalu, S.-R., M. Maijala, M. Maunumäki, M. Maunula, and M.A. Shiweda. 2025. Ecological and Economic Sustainability in Namibian and Finnish Pre-Service Language Teachers' Reflections. In *Generating Sustainable Futures Through Teacher Education,* ed. H. Harju-Luukkainen, S. Garvis, J. Kangas, J. Marôco, M. Maunula, and M. Maunumäki, 223–39. Rethinking Higher Education. Singapore: Springer Nature Singapore, 2025. https://doi.org/10.1007/978-981-96-3329-6_14.

Laine, P., S.-R. Kuusalu, M. Maijala, and M. Mutta. 2022. Kestävä kehitys kieltenopetuksessa: Kieliaineiden opettajaopiskelijoiden käsityksiä kestävyysaiheista kielten oppitunneilla. *AFinLAn Vuosikirja* 79 (December): 109–32. https://doi.org/10.30661/afinlavk.114529.

Litre, G., F. Hirsch, P. Caron, A. Andrason, N. Bonnardel, V. Fointiat, W.O. Nekoto, et al. 2022. Participatory detection of language barriers towards multilingual sustainability (ies) in Africa. *Sustainability* 14, no. 13 (July 4): 8133. https://doi.org/10.3390/su14138133.

Lähdemäki, J. 2019. Case study: The Finnish National Curriculum 2016–A co-created national education policy. In *Sustainability, human well-being, and the future of education,* ed. J.W. Cook, 397–422. Cham: Springer International Publishing. https://doi.org/10.1007/978-3-319-78580-6_13.

Lavonen, J. 2018. Educating professional teachers in Finland through the continuous improvement of teacher education programmes. In *Contemporary pedagogies in teacher education and development*, ed. Y. Weinberger and Z. Libman, 3–22. London: IntechOpen.

Maijala, M. 2020. Pre-service language teachers' reflections and initial experiences on the use of textbooks in classroom practice. *Journal of Language Teaching and Research* 11, no. 2 (March 1): 157. https://doi.org/10.17507/jltr.1102.03.

Maijala, M., and M. Tammenga-Helmantel. 2019. Grammar exercises in Dutch, Finnish and global textbooks for teaching German as a foreign language. *The Language Learning Journal* 47, no. 5 (October 20): 537–557. https://doi.org/10.1080/09571736.2017.1309449.

Maijala, M., A. Roiha, and M. Patzelt. 2023a. Vieraiden kielten aineenopettajaopiskelijoiden sähköposteja tulevaisuudesta: Tarkastelussa teknologiset oppimisympäristöt. *Ainedidaktiikka* (September 14). https://doi.org/10.23988/ad.127899.

Maijala, M., L.M. Heikkola, S.-R. Kuusalu, P. Laine, M. Mutta, and K. Mäntylä. 2023b. Pre-service language teachers' perceptions of sustainability and its implementation in language teaching. *Language Teaching Research* (May 13). https://doi.org/10.1177/13621688231170682.

Mare, A., E. Woyo, and E.M. Amadhila. 2023. *Teaching and learning with digital technologies in higher education institutions in Africa. Case studies from a pandemic context*, 1st ed. Oxford: Routledge.

MAWF (Ministry of Agriculture, Water, and Forestry). 2008. *National textbook policy. National planning commission of Namibia.* Available from: https://www.npc.gov.na/publicpolicy/textbook-policy/. Accessed September 26, 2024.

Mbangula, D.K. 2024. Exploring the integration of ICT in public sector management in Namibia: In *Advances in information quality and management*, 6th ed., ed. M. Khosrow-Pour, D.B.A., 1–13. IGI Global. https://doi.org/10.4018/978-1-6684-7366-5.ch039.

Ministry of Education. 2010. *The national curriculum for basic education.* Okahandja, Namibia: National Education Institute Department.

Ministry of Education and Culture. 2022. *Teacher education development programme 2022–2026.* Available from: https://valtioneuvosto.fi/-/1410845/opettajankoulutuksen-kehittamisohjelma-uudistettiin-vastaamaan-2020-luvun-tarpeita?languageId=en_US. Accessed June 26, 2024.

Ministry of Education and Culture. 2023. *Policies for the digitalisation of education and training until 2027.* Helsinki: Ministry of Education and Culture.

MEFT. Ministry of Environment, Forestry and Tourism. 2019. *National Environmental Education and Education for Sustainable Development Policy.*

Available from: https://www.meft.gov.na/files/downloads/a27_NaDEET%20EE%20Policy.pdf. Accessed May 23, 2024.

Mpofu-Hamadziripi, N., F. Rauch, and M. Dulle. 2022. Transforming curricula in higher education: Description of two perspectives from the Global South and the Global North. *Frontiers in Education* 7 (February 25): 775406. https://doi.org/10.3389/feduc.2022.775406.

NIED. 2016. *Summary of conference proceedings and recommendations from the national workshop on African languages in basic education held at NIED in Okahandja*, 18–22 September 2016.

Ngololo, E.N., and H. Nekongo-Nielsen. 2017. Teachers' views on the implementation of the English language proficiency programme in Namibia. *International Education Studies* 10, no. 11 (October 29): 155. https://doi.org/10.5539/ies.v10n11p155.

Norro, S. 2021. Namibian teachers' beliefs about medium of instruction and language education policy implementation. *Language Matters* 52, no. 3 (September 2): 45–71. https://doi.org/10.1080/10228195.2021.1951334.

Norro, S. 2022. Factors affecting language policy choices in the multilingual context of Namibia: English as the official language and medium of instruction. *Apples—Journal of Applied Language Studies* (February 16). https://doi.org/10.47862/apples.107212.

OAJ (The Trade Union of Education in Finland). 2021. *Dramatic shift: Six in ten Finnish teachers are now considering career change.* Available from: https://www.oaj.fi/en/news/news-and-press-releases/2021/alanvaihtokysely-09-21/. Accessed April 13, 2024.

OECD. 2023. *Country digital education ecosystems and governance: A companion to digital education outlook 2023.* Paris: OECD Publishing.

O'Sullivan, M.C. 2001. Communicative approaches to teaching English in Namibia: The issue of transfer of Western approaches to developing countries. Les approches communicatives de l'enseignement de l'Anglais en Namibie: La question du transfert des méthodes occidentales aux pays en voie de développement. Enfoques comunicativos de la enseñanza del Inglés en Namibia: El tema de la transmisión de enfoques occidentales a los países en vías de desarrollo. *International Journal of Early Years Education* 9, no. 1 (March): 51–61. https://doi.org/10.1080/713670675.

Patton, M.Q. 2015. *Qualitative research & evaluation methods: Integrating theory and practice*, 4th ed. Thousand Oaks, CA: SAGE.

Paronen, P., and O. Lappi. 2018. *Finnish teachers and principals in figures. Reports and Surveys* 4. Helsinki: Finnish National Agency for Education.

Pegalajar-Palomino, M.D.C., A. Burgos-García, and E. Martinez-Valdivia. 2021. What does education for sustainable development offer in initial teacher training? A systematic review. *Journal of Teacher Education for Sustainability* 23, no. 1 (June 1): 99–114. https://doi.org/10.2478/jtes-2021-0008.

Pollari, P., O. Salo, and K. Koski. 2018. In teachers we trust—The Finnish way to teach and learn. *Inquiry in Education* 10, no. 1. Available from: https://digitalcommons.nl.edu/ie/vol10/iss1/4/. Accessed November 6, 2024.

Repo, E. 2023. Towards language-aware pedagogy? Experiences of students in multilingual Finnish schools. *Language and Education* 37, no. 4 (July 4): 460–482. https://doi.org/10.1080/09500782.2022.2116985.

Republic of Namibia. 2008. *Namibia National Textbook Policy*. Windhoek: National Planning Commission of Namibia.

Richards, J.C. 2014. The ELT textbook. In *International perspectives on materials in ELT*, ed. S. Garton and K. Graves, 19–36. Basingstoke, UK: Palgrave Macmillan.

Ruohotie-Lyhty, M., and A. Pitkänen-Huhta. 2022. Status versus nature of work: Pre-service language teachers envisioning their future profession. *European Journal of Teacher Education* 45, no. 2 (March 15): 193–212. https://doi.org/10.1080/02619768.2020.1788535.

Sahlberg, P. 2018. PISA in Finland: An education miracle or an obstacle to change? *Center for Educational Policy Studies Journal* 1, no. 3 (January 18): 119–140. https://doi.org/10.26529/cepsj.418.

Säntti, J., M. Puustinen, and P. Hansen. 2023. Unmentioned challenges of Finnish teacher education: Decontextualisation, scientification and the rhetoric of the research-based agenda. In *Finland's famous education system*, ed. M. Thrupp, P. Seppänen, J. Kauko, and S. Kosunen. Singapore: Springer. https://doi.org/10.1007/978-981-19-8241-5_7.

Singh, S. 2020. From Global North-South divide to sustainability: Shifting policy frameworks for international development and education. *International and Multidisciplinary Journal of Social Sciences* 9, no. 1 (March 30): 76. https://doi.org/10.17583/rimcis.2020.4923.

Shava, G.N., D. Nkengebeza, N. Phuthi, K.D. Mpuang, and J. Alava. 2021. Achieving Sustainable Development Goal (SDG) 4 on quality in education: challenges and opportunities, experiences from Southern African (SADC) universities. *International Journal of Research and Innovation in Social Science (IJRISS)* 5, no. 7 (July): 559–573.

Sommier, M., Y. Wang, and A. Vasques. 2022. Transformative, interdisciplinary and intercultural learning for developing HEI students' sustainability-oriented competences: A case study. *Environment, Development and Sustainability* (March 1). https://doi.org/10.1007/s10668-022-02208-7.

Spiropoulou, D., T. Antonakaki, S. Kontaxaki, and S. Bouras. 2007. Primary teachers' literacy and attitudes on education for sustainable development. *Journal of Science Education and Technology* 16, no. 5 (October 15): 443–450. https://doi.org/10.1007/s10956-007-9061-7.

Steadman, A., H. Kayi-Aydar, and S.M. Vogel. 2018. From college composition to ESL: Negotiating professional identities, new understandings, and conflicting pedagogies. *System* 76 (August): 38–48. https://doi.org/10.1016/j.system.2018.04.013.

TENK. Finnish National Board on Research Integrity. 2023. *Responsible conduct of research (RCR).* Available from: https://tenk.fi/en/research-misconduct/responsible-conduct-research-rcr. Accessed June 6, 2024.

Tirri, K. 2014. The last 40 years in Finnish teacher education. *Journal of Education for Teaching* (September 10): 1–10. https://doi.org/10.1080/02607476.2014.956545.

Tomlinson, B. 2012. Materials development for language learning and teaching. *Language Teaching* 45, no. 2 (April): 143–179. https://doi.org/10.1017/S0261444811000528.

Toom, A., and J. Husu. 2021. Analyzing practice, research, and accountability turns in Finnish academic teacher education. In *Teacher Education Policy and Research,* ed. D. Mayer, 73–83. Singapore: Springer Nature Singapore. https://doi.org/10.1007/978-981-16-3775-9_6.

Uitto, A., and S. Saloranta. 2017. Subject teachers as educators for sustainability: A survey study. *Education Sciences* 7, no. 1 (January 4): 8. https://doi.org/10.3390/educsci7010008.

UN. 2022. United Nations Sustainable Development Group. (2022, September 8). *The power of coding: UN teams partner up to close the digital literacy gap in Namibia and beyond*. United Nations. https://unsdg.un.org/latest/stories/power-coding-un-teams-partner-close-digital-literacy-gap-namibia-and-beyond

UNESCO. 2020. *Education for Sustainable Development: A roadmap.* France: UNESCO.

United Nations. 2015. *Transforming our world: The 2030 agenda for sustainable development.* Available from: https://sdgs.un.org/2030agenda. Accessed June 26, 2024.

Varpanen, J., A. Laherto, J. Hilppö, and T. Ukkonen-Mikkola. 2022. Teacher agency and futures thinking. *Education Sciences* 12, no. 3 (March 3): 177. https://doi.org/10.3390/educsci12030177.

Weinberg, A.E., C.D. Trott, W. Wakefield, E.G. Merritt, and L. Archambault. 2020. Looking inward, outward, and forward: Exploring the process of transformative learning in teacher education for a sustainable future. *Sustainability Science* 15, no. 6 (November): 1767–1787. https://doi.org/10.1007/s11625-020-00831-9.

6

Sustainability Content in German Language Coursebooks in Europe

Minna Maijala, Mareen Patzelt, and Friederike Delius

Introduction

The concept of sustainability has emerged as a significant topic of discussion across a range of academic disciplines, and it is now receiving increasing attention within the field of language education and research (see Chapter 1 in this volume). Many countries, including those in Europe, have adopted sustainability-related content in their school

M. Maijala (✉) · M. Patzelt
School of Languages and Translation Studies, University of Turku, Turku, Finland
e-mail: minmai@utu.fi

M. Patzelt
Åbo Akademi University, Turku, Finland

F. Delius
University of Marburg, Marburg, Germany

University of Porto, Porto, Portugal

M. Maijala et al. (eds.), *Transformative Language Teaching for Sustainability*,
https://doi.org/10.1007/978-3-031-85493-4_6

curricula (Gericke et al. 2020), mostly due to UNESCO's (2018) Agenda 2030 call for action. The key to reaching Sustainable Development Goals (SDGs) is Education for Sustainable Development (ESD), which refers to pedagogical principles and learning processes that promote awareness and competencies related to SDGs and enable change (e.g., Eilam and Trop 2010). Promoting ESD in the teaching of all subjects requires teachers to learn transformative competencies, which include systems thinking. Systems thinking is a holistic and integrated approach to understanding complex phenomena, which involves studying these phenomena from multiple perspectives simultaneously to obtain a comprehensive understanding of how different processes not only work but also how they affect the larger system (Lewis et al. 2014; Sinakou et al. 2019). Other key competencies for advancing sustainability transformation include developing and implementing strategies that enable actions that promote sustainability (Redman and Wiek 2021). These strategies should be implemented in collaboration with one another (Frisk and Larson 2011). In language learning, a range of transferable skills can be learned to promote ESD (de la Fuente 2021; Maijala et al. 2024), such as communication skills, collaboration, critical thinking and problem solving, systems thinking, reflection as well as active participation in decision-making processes and in community involvement (e.g., Riess et al. 2022). Several integrative and learner-oriented methods that already belong to language teaching traditions and are widely used in language classrooms (Maijala et al. 2024) can, to some extent, be found in language textbooks (e.g., Lasekan et al. 2023).

Previous studies have shown that the implementation of ESD is significantly influenced by disciplinary traditions among subject teachers (Sund et al. 2020). Even when addressing identical topics, different subject teachers approach curricular content in distinct ways (Gericke et al. 2018). Educational resources, particularly textbooks, play a significant role in shaping curriculum and classroom discourse (Guerrettaz & Johnston 2013; Tomlinson 2012). Language textbooks, in particular, can be viewed as cultural artefacts that reflect the context of their use (Gray 2010; Harwood 2013). These materials often contribute to what is known as the *hidden curriculum* (e.g., J. Lee 2014b; Vu and Pham

2022), especially when educators fail to recognise the underlying motivations behind textbook content (Hahl et al. 2015). It is worth noting that many language teachers and pre-service teachers tend to regard textbooks as the curriculum itself (Carabantes and Paran 2022). Therefore, the presence of sustainability content in language textbooks directly impacts the extent and manner in which ESD is implemented in language classrooms. However, there is a noticeable gap in research concerning the role of ESD in language textbooks. International comparisons are necessary to gain a more comprehensive understanding of how sustainability is incorporated into these educational materials. This disparity highlights the need for more diverse and inclusive research in the field of language education and sustainability, especially in middle income countries and the Global South.

Against this background, the aim of this chapter is to determine how sustainability is thematised in various German coursebooks for learners aged approximately 15–17 years learning German at the A1–B2 level on the CEFR scale. The coursebooks analysed are from the UK, Denmark, Finland, the Netherlands, and Norway. In the literature review, we discuss the role of textbooks in promoting ESD in general and the previous research literature on SDG-related content in textbooks. In the analysis, both texts and exercises with SDG-related content of the coursebooks are counted and classified and then analysed qualitatively. We then present our findings and reflect on their meaning and implications for language education.

The Role of Language Textbooks in Promoting Education for Sustainable Development

In the language education research literature, we can find some distinctive features of language teaching and teachers (Borg 2006) that influence the extent to which and how ESD can be implemented in language teaching (Maijala et al. 2024). The subject matter of language teaching is difficult to define because communication in the target language is the goal of the teaching and the way to reach this goal (e.g., Hammadou and Bernhardt 1987). Additionally, in language textbooks, inputs to intercultural learning are integrated into communicative activities that might, for example, foster awareness of global citizenship and responsibility. For instance, students may engage in cross-cultural interactions with peers from other countries or undertake linguistic challenges, such as identifying street signs in a variety of languages (Risager 2022). Language textbooks frequently incorporate communicative activities that address real-life problems (e.g., H. Y. Lee 2014a), and which could serve as opportunities for reflective thinking in language teaching, thus aligning with the principles of ESD (Sinakou et al. 2019; Maijala et al. 2024). The learning content in language textbooks typically encompasses exercises in a range of linguistic domains, including listening, speaking, writing, vocabulary, grammar, and pronunciation (Littlejohn 2022). Exercise types tend to be mostly deductive, that is, following the so-called PPP (*presentation-practice-production*) approach (Tomlinson 2022), which consists mostly of completion and matching, such as blank-filling exercises (Maijala and Tammenga-Helmantel 2019). The prevailing approach to language teaching in textbooks can be characterised by an emphasis on a traditional methodology, with a focus on controlled practice and a relatively limited emphasis on communication (Tomlinson and Masuhara 2013). The open-ended discovery approach has not been a common practice in language textbooks (Tomlinson 2022).

In the European context, the Common European Framework of Reference for Languages (CEFR; Council of Europe 2001, 2020) provides guidelines for planning curricula and textbook contents and for choosing topics in language teaching (Byram 2022). Besides textbooks, the teachers' role is crucial in integrating sustainability topics into classroom practices (Wiek et al. 2015) because their pedagogical choices influence the students' learning experience (Öhman and Sund 2021). According to several studies (e.g., Tomlinson 2012; Rathert and Cabaroğlu 2021), language teachers tend to rely on the content presented in textbooks when designing their lessons. For instance, in the Finnish context, textbooks are used extensively in language teaching (e.g., Harjanne et al. 2017) and language teachers choose topics for their lessons mostly from textbooks (e.g., Hahl 2020). Textbooks guide teachers' lesson planning as well as content and activity selection (e.g.,

Harwood 2010; McGrath 2006; Richards 2014). Accordingly, although the selection and use of teaching materials vary in different educational contexts, textbooks remain the most common source for language teaching and, together with other teaching materials, have a great influence on which issues are addressed in language classrooms (e.g., Guerrettaz and Johnston 2013). Consequently, to promote the integration of ESD in language teaching globally, it is essential to place a greater focus on the specific topics of sustainability and activities related to them, as well as the interconnections between them. The content in language textbooks is mostly focused on linguistic and cultural aspects (e.g., Weninger and Kiss 2013), that is, topics related to the cultural and social dimensions of sustainability (e.g., Mohammadnia and Moghadam 2019). Textbooks also tend to show an ideologically stereotyped view of the world, favouring middle-class lifestyles and consumerism (e.g., Gray 2010) over ideological and hegemonic elements, which are perceived as potentially risky in language teaching settings (Ulum and Köksal 2021).

By guiding the discourse and activities (Guerrettaz and Johnston 2013; Littlejohn 2022), language coursebooks play a pivotal role in implementing ESD in language classrooms. Therefore, language textbooks could facilitate the promotion of ESD by incorporating content and activities aligned with SDGs (Jodoin and Singer 2019). Accordingly, the selection of content and exercises in textbooks impacts how ESD is implemented in language classrooms.

Previous Research Literature on Sustainability Content in Textbooks

Some existing research has examined SDG-related content in textbooks for subjects such as geography and biology (e.g., Biström and Lundström 2021), social science (e.g., Satokangas and Mikander 2023), and primary education (e.g., Vásquez et al. 2021). Research on implementing ESD in language textbooks has not yet received notable attention. However, in the early 2000s, researchers had already conducted a few ecolinguistic studies of textbooks for English as a second or foreign language. Jacobs and Goatly (2000) examined how ecological issues are treated in English coursebooks and concluded that teaching materials do not explicitly encourage participation in environmental protection. Using a critical discourse framework, Stibbe (2004) analysed how twenty-six Japanese English as a foreign language (EFL) textbooks thematised environmental education. The findings showed that the textbooks mainly presented technical solutions to environmental problems. Although the environmental discourse in textbooks was mostly based on 'shallow environmentalism', there were a few examples of a deeper understanding of ecology. Xiong (2014) conducted an eco-critical discourse analysis of Chinese secondary school EFL textbooks. The findings indicated that the environmental topics in the textbooks raised students' ecological awareness but failed to encourage their active participation and transformational practices.

Recent studies have examined how EFL textbooks treat sustainability content in different educational contexts. Several studies reported that ESD-related content was lacking (e.g., review by Mochizuki 2019) and that exercises in language textbooks varied in the extent of promoting change agency (Cristovão et al. 2022; Zahoor and Janjua 2020). However, other studies have presented a positive picture of ESD implementation in EFL textbooks. For example, Lasekan and others (2023) examined SDG-related themes in EFL textbooks in the Chilean context and identified grammatical tasks that had the potential to foster students' sustainability competencies. The textbooks addressed almost all of the seventeen SDGs, except SDGs 1 (water) and 6 (poverty). The grammatical tasks could also enhance sustainability competencies. Grammar exercises can, for example, promote systems thinking skills by analysing grammatical rules. In another recent study, Cristovão and others (2022) examined environmental discourse in Brazilian EFL textbooks used in public secondary schools. The findings showed that the textbooks raised students' eco-awareness by including environmental issues. However, they did not provide students with educational experiences that would have promoted their critical environmental literacy or motivated them to engage actively. In the Iranian context, Mohammadnia and Moghadam (2019) analysed an English textbook series. Their findings revealed that sustainability topics were presented in regionally used Iranian textbooks to a noticeable extent, but the topics were distributed across different volumes. In Pakistan, Zahoor and Janjua (2020) investigated green content in English textbooks at the primary level and found that representations of nature lacked eco-pedagogical values. However, the Pakistani textbooks included practical activities to promote children's skills and participation in resolving environmental crises.

The number of studies examining the incorporation of SDG-related content into textbooks of languages other than English is still relatively limited. Risager (2021) analysed Danish language textbooks for English, German, French, and Spanish learners aged 13 and above focusing on the representation of different countries, key global issues, cultural and social identities, aspects of world history, and connections between different regions. The analysis of key global issues included topics such as climate change, inequality, and poverty. The findings showed that environmental problems were explicitly addressed, and one textbook even encouraged learners to take an active role. These problems were often addressed through the self-descriptions of young textbook protagonists. However, the representations of environmental problems were rather fragmented, and the textbooks did not address these issues in a comprehensive manner. Curdt-Christiansen (2021) explored environmental literacy topics in Chinese language textbooks used in elementary school education. The analysis focused on texts and topics introducing environmental literacy and the ways in which Chinese children are positioned as environmental actors. The content analysis focused on how different text features (text types, lexical choices, grammatical elements, and generic structures) were used to reflect environmental topics and promote children's awareness of the environment. However, the findings showed that the content and topics were limited. The books overemphasised consumer behaviour as a reactive strategy rather than a proactive engagement to protect the environment. Children were mostly positioned as actors whose behaviour damaged the environment. The books also provided limited content on ways to develop the ability to investigate and evaluate environmental problems and take actions to solve them.

To sum up, integrating SDGs and ESD into textbooks across all subjects, including language textbooks, presents a significant challenge worldwide. When integrating SDG-related content and activities, textbooks have difficulties creating relationships between developing a holistic understanding of sustainability dimensions and enhancing students' competencies for action. Furthermore, incorporating reflections on textbook content with the teacher and other learners, as recommended by research as an effective method for achieving SDGs, could also prove to be a challenge (Micalay-Hurtado and Poole 2022; Stibbe 2021). Overall, previous studies on this topic remain incomplete because they have been conducted in different educational settings. In addition, most studies on sustainability content in language textbooks have so far focused on teaching English (e.g., Katunich and Goulah 2020), and a systematic understanding of how ESD-related content is integrated into language textbooks other than English is lacking. The present study therefore contributes by investigating how ESD content is integrated into German language textbooks used in several European countries.

Study Design

We chose German language coursebooks as research material to assess how sustainability content areas are included in various textbooks on German as a foreign language based on the following research questions (RQ):

RQ1 What sustainability content areas are included in the coursebooks?
RQ2: How do these coursebooks cover sustainability content areas?

The present study examines coursebooks from the UK, Finland, the Netherlands, Norway, and Sweden. In all countries included in this study, the curriculum for language teaching and learning is strongly influenced by the CEFR (Council of Europe 2001, 2020). Although the coursebooks analysed align with the national curricula, it is not possible to provide details of each country's SDG-related objectives due to word limit constraints. For instance, in Finnish school curricula, sustainability education is considered at all levels of the education system. This can be seen, for example, in the cross-cutting objectives of the Finnish general upper secondary education curriculum. Many of these objectives are also incorporated into language teaching, including skills such as interaction, ethics, and global and cultural competence (EDUFI 2019).

Materials

The study focuses on locally used German coursebooks for beginners aged approximately 15–17 years who are learning German (A1–B2 on the CEFR scale) available in 2024 in Denmark, Finland, the UK, the Netherlands, and Norway. All books were produced and used in the respective countries. To analyse the sample within the time available for the study, we included up to two coursebooks from each country. Altogether, our research data comprise eleven coursebooks, including texts and exercises. We used the abbreviations in brackets to designate the analysed coursebooks in all of the examples provided in this chapter. Appendix 1 provides detailed information and descriptions of the analysed books (see Table 6.1). The coursebooks do not represent *language education* in the respective countries, as this study does not cover all German language coursebooks published in the respective countries.

Table 6.1 Coursebooks and their abbreviations

Coursebook title	Coursebook abbreviation
Klasse! 2	N 1
Weiter geht's	N 2
Auf uns. Grundbog til tysk i gymnasiet	DK 1
Los geht's!	DK 2
Na klar! Duis voor de onderbouw 3	NL 1
Neue Kontakte, Deutschbuch A	NL 2a
Neue Kontakte, Deutschbuch A	NL 2b
AQA GCSE	GB 1
Stimmt! Edexcel GCSE (9–1)	GB 2
Magazin.de 1	FIN 1
Magazin.de 2	FIN 2

Methods

The data for this study were collected by all authors reading the coursebooks thoroughly. Any pages that included texts and exercises around sustainability themes were photographed, stored in NVivo (Cohen et al. 2018) and tagged according to the name of the textbook and the current page. Altogether, we obtained 571 pictures of pages with exercises and/or texts with sustainability content from the eleven coursebooks (Table 6.1). We thus analysed an average of fifty-two pages per textbook, with a maximum of 163 per textbook (N 2) and a minimum of fourteen pages per textbook (FIN 1).

We used a qualitative content analysis method to interpret the dynamic interactions between the data and theory (Corbin and Strauss 2014). Although textbooks for language learning traditionally include numerous visual elements, such as illustrations and symbols (Weninger and Kiss 2013), we focused solely on the written texts and the themes they displayed. For the listening exercises, we used only the information retrieved from the written part of the exercise in the coursebooks.

For coding, we identified five subthemes for each of the four dimensions of sustainability (see Table 6.2). Subthemes were defined based on UNESCO's Earth Charter (2000), the Roadmap for Implementing the Global Action Program on ESD (UNESCO 2020), and pre-existing theory and research literature (Maijala et al. 2023; Mohammadnia and Moghadam 2019). We also had predefined examples of topics belonging to each subtheme (see Table 6.2). These topics were partly predefined from theory (Maijala et al. 2023; Mohammadnia and Moghadam 2019) and partly emerged from the material during the analysis.

For each subtheme, we identified six different codes (see Table 6.3), that is, three categories each for texts (T) and exercises (E). This means that, within both of these categories, we coded three different subcategories (1, 2, 3) according to the way in which a sustainability topic was part of the text or exercise in question. Within categories T and E, texts and exercises were coded into one subcategory, from one to three, depending on how explicitly and holistically the texts and exercises featured sustainability topics (Table 6.3). Texts coded into subcategory T3 explicitly inspired action in the reader or—for example, through

Table 6.2 Subthemes of ecological, social, cultural, and economic sustainability coded per each sustainability dimension. Each subtheme (in bold) is further coded into subcategories T1–T3 and E1–E3 (explained further in Table 6.3)

Ecological sustainability	Social sustainability	Cultural sustainability	Economic sustainability
Reducing and managing waste	**Democracy and peace**	**Societal justice and well-being**	**Reducing emissions**
Promoting recycling (second-hand shopping) Reducing air pollutants Reducing litter	Promoting world peace Strengthening democracy Ensuring possibility to influence activism Reflecting on historical events relevant to peace Civil courage	Promoting societal justice Promoting societal well-being Activism Charity, volunteering	Reducing industrial emissions Reducing agricultural emissions Reducing traffic emissions Responsible consumption (fashion, entertainment) Circular economy
Mitigating climate change	**Human rights and equality**	**Preserving and supporting culture in all its forms**	**Energy, technology, and use of natural resources**
Activism Responsible ways of living Ways to talk about climate	Promoting gender equality Securing minority rights Securing human rights Reflections about family Preventing discrimination and racism Activism Reflection on friendship Reflection on differences and similarities Accessibility	Preserving cultural heritage Supporting minority cultures Supporting different forms of art as part of culture Spiritual practices	Developing low-carbon energy solutions Saving energy Limiting the use of natural resources Developing environmental-friendly technology
Conserving nature and wildlife	**Securing health & healthy living conditions**	**Overall security and legal services**	**Favouring ethically sustainable products**

(continued)

Table 6.2 (continued)

Ecological sustainability	Social sustainability	Cultural sustainability	Economic sustainability
Securing biodiversity Nature conservation Wetland conservation Preventing environmental chemicalisation	Preventing bullying Maintaining health (e.g., through sports and nutrition) Digital well-being, technology Awareness, responsible technology use Fleeing unhealthy and insecure living conditions	Promoting equal access to legal services Enhancing overall security Volunteering (for example fire brigade)	Learning about food traditions Learning about food industry Favouring vegetable-based food Fair trade
Securing ecosystem services	**Securing a social security system**	**Favouring responsible travel**	**Enhancing waste recovery**
Preventing natural disasters Securing water supply	Preventing poverty Preventing homelessness	Favouring train travel over planes Favouring car-pooling Favouring public transportation Focus on natural aspects of travel destinations	Infrastructure development
Protecting animal rights	**Securing employment**	**Preserving linguistic diversity**	**Reducing indebtedness**
Choosing ethically produced animal products Activism	Educational (in)equality Reflecting on career paths Activism	Regional varieties of languages Linguistic diversity within target language region Reflecting on differences and similarities between languages	Inflation Economic growth

Table 6.3 Coding principles for subcategories of texts and exercises

T1/ E1	Sustainability mentioned implicitly in singular words and phrases, not main topic for T/E
T2/ E2	Clear thematic focus on sustainability in T/E, sustainability topic mentioned in headline for the entire T/E
T3/ E3	*Advanced subcategory* T/E has explicit focus on sustainability and promotes holistic understanding of sustainability as well as an action-oriented approach beyond the textbook and/or learning situation

giving background information or making comparisons—displayed a direct link to the learner's own reality or a holistic understanding of sustainable development (Sinakou et al. 2019). In the analysis, T3/E3 was designated as an *advanced subcategory*.

During the initial coding process, the researchers made observations independently. Then, we consolidated our views, made modifications and specifications to the initial codes, and came to a common understanding (Patton 2015) for the final coding structure visible in Table 6.2. Despite this common understanding, the borders between some codes may still be perceived as fluent. This is one of the reasons we did not take a quantitative approach in this research. Another reason is that some texts and exercises spread over more than one page received a separate code for each part, and if the coursebooks included a text and an exercise specifically for this text, the two were coded separately into the categories T and E. The results from this study are thus qualitative in nature, as the coding method does not allow for a quantitative analysis overall.

In general, the texts and exercises we analysed displayed more than one of the subthemes for sustainable development. If there was a dominant topic, we coded the entire exercise or text into this topic in subcategory 2 or 3 and parts of the exercise or text that mentioned other topics on the side in subcategory 1. The coursebooks differed greatly from one another regarding the amount of sustainability content they contained and the way in which they addressed sustainability topics. Therefore, the results provide an overview of what kind of sustainability content can currently be encountered in textbooks, not so much comparing the textbooks of different countries and/or learner levels with one another.

We want to note that the differences in the language level of the coursebooks may influence the consideration of the SDG-related content. The content of the books is also influenced by diverse learning traditions and the learners' first languages. For instance, in countries where the spoken language is similar to German (e.g., Dutch), learners might progress further in their German studies than in countries where the spoken language is more distant from German (e.g., Finnish). There are also differences in the traditions of how textbooks are structured (Harwood 2013). The status of the German language also differs from country to country. In Western European countries and Nordic countries, the number of German learners is declining (Auswärtiges Amt 2020). Consequently, the language level of the books varies and represents a limitation for this study. To gain a clearer picture of the coursebook content, illustrative examples are given (the examples were translated into English by the authors).

Results

Sustainability Content Areas in Coursebooks

To answer the first research question regarding what sustainability content areas were included in the coursebooks, we analysed which of the four sustainability dimensions were identifiable in the material overall and in each textbook (see Table 6.4). We identified all four sustainability dimensions across the material. However, only social sustainability was featured in all eleven coursebooks. Cultural sustainability was found in ten coursebooks, economic sustainability in seven, and ecological sustainability in six coursebooks.

To address the second research question, that is, the way in which coursebooks cover different sustainability content areas, we extracted the texts and exercises in question and coded them into the *advanced subcategory* T/E 3 (see Table 6.3). The number of coursebooks featuring exercises coded into the third category is larger than the number of texts coded into the third category. Two coursebooks included neither texts nor exercises in this subcategory (FIN 1 and FIN 2), and three books

Table 6.4 Number of pages in coursebooks with sustainability content and the inclusion of sustainability dimensions in coursebooks

Coursebook	Number of pages with sustainability content (texts and/or exercises)	Dimensions of sustainability included			
		Ecological sustainability	Social sustainability	Cultural sustainability	Economic sustainability
N2	163	X	X	X	X
NL 1	65		X	X	X
DK 1	59		X	X	
GB 1	53	X	X	X	X
NL 2b	46	X	X	X	X
N1	43	X	X	X	X
GB 2	41	X	X	X	X
DK2	36		X		
NL 2a	33	X	X	X	
FIN 2	18		X	X	X
FIN 1	14		X	X	

Table 6.5 Texts and exercises coded into the *advanced subcategory*. Subthemes that were covered in both texts and exercises are marked with bold font

Sustainability dimensions	Subthemes
Ecological sustainability	**Conserving nature and wildlife** **Mitigating climate change** **Reducing and managing waste**
Social sustainability	**Democracy and peace** **Human rights and equality** Securing a social security system Securing employment Securing health and healthy living conditions
Cultural sustainability	Favouring responsible travel **Preserving and supporting culture in all its forms** **Societal justice and well-being**
Economic sustainability	**Energy, technology, and use of natural resources** Enhancing waste recovery Favouring ethically sustainable products **Reducing emissions**

included both texts and exercises in the third subcategory (NL 2b, GB 2, N 2). Four coursebooks contained exercises within this subcategory but no texts (GB 1, DK 1, DK 2, NL 1, NL 2a).

Texts and exercises in the *advanced subcategory* focused on various subthemes from all four sustainability dimensions. Most of the subthemes mentioned in Table 6.2 could be identified both in texts and exercises (marked with bold font in Table 6.5), others only in exercises.

Several texts and exercises included more than one subtheme in most of the coursebooks. Sometimes, these also belonged to more than one dimension of sustainability, thus connecting various topics to one another and aiming to paint a bigger picture of what the text was about. One example is an exercise about prejudices humans tend to hold. The exercise linked reflections on minorities (social sustainability) and identity with aspects of preserving both individual and shared cultures (cultural sustainability) (Example 1).

(1) What comes to mind?

What is positive about a diverse society?

> What do racism and anti-racism mean to you?
> What examples of racism do you see in your community?
> How can we combat prejudice and racism?
> ['An unsympathetic dialogue': Dialogue in which one person wonders about another person's German skills and their origin and assumes this person could not possibly be German, as they supposedly do not look German because of their Ghanaian ancestry]
> What is it about?
> What is the topic of conversation in this dialogue?
> Where are the people from?
> Why does one person think that the other person is not German?
> What do you think of the ending?
> Write something!
> Rewrite the unsympathetic dialogue so that it becomes sympathetic.
> (N 2, p. 180)

Another example is a text connecting its main topic of conserving nature and wildlife to reducing and managing waste (both ecological sustainability) and favouring ethically sustainable products (economic sustainability). The text's heading is 'Watt ein Meeresmüll!' (NL 2b, pp. 122–123), referring, through wordplay, to the mudflats in Northern Germany and the pollution of the North Sea. It is about a school class taking part in a project to reduce plastic waste in the North Sea through concrete actions, such as collecting waste on the beach, as well as conscious consumption that can help to avoid releasing microplastic into the environment. The text addresses various possible positive consequences of these actions, such as the conservation of nature and wildlife, as well as human health.

Sustainability Content in the *Advanced Subcategory*

To answer the second research question, that is, to investigate more closely how coursebooks cover sustainability, we first categorised the texts and exercises according to the three subcategories (T1, T2, T3 and E1, E2, E3) explained in Table 6.3. To assess how coursebooks integrated sustainability content in a way that promoted its holistic understanding and enabled students to take action, we defined the characteristics for those tasks and texts coded into the third subcategory in the material.

Texts within the *advanced subcategory* gave concrete examples of taking action for sustainability that was directly relatable to the students' own lives, either because the texts showed how youth become active in sustainable development or because the text explicitly emphasised how the topic is relevant to students and part of a bigger picture in general. Within the dimension of cultural sustainability, for instance, texts showed examples of youth volunteering for UNICEF or provided historical and cultural background information on the multicultural society in Germany nowadays. Texts coded into social sustainability, for example, pointed out in what ways Anne Frank's everyday life was similar and different from the one the students live nowadays and showed that integrating migrants into a society is a task for everybody who is part of that society. A text coded into the third subcategory in the dimension of ecological sustainability shows a specific example of an ecological volunteer gap year for youth in Germany. One text coded into the third subcategory for financial sustainability explained innovative and cost-efficient ways in which a self-sufficient village in Germany produces its own energy from renewable resources.

Exercises coded into the *advanced subcategory* typically included creative writing, discussions where students take up different perspectives or reflect for themselves and express their opinion, and additional research on a certain topic that helped students understand the context better. The following four exercises are examples of how exercises coded into subcategory 3 appeared in the coursebooks.

In one exercise coded into the subtheme of the cultural dimension *preserving and supporting culture in all its forms*, students are asked to work in pairs and discuss whether to have graffiti as pieces of art around their school or around their town (Example 2).

(2) Work in pairs and choose one of the three situations:

1. A large, colourful 'piece' appeared on the wall of your house last night. Discuss whether it should stay or be removed. Person 1 is a tenant (and wants to keep the graffiti). Person 2 is the janitor

(and wants to remove the graffiti). Present the role-play to the class.

2. There is an empty wall in your schoolyard. Should you invite a graffiti artist to paint it or should it stay white? Person 1 is a pupil. Person 2 is the principal. Present the role-play to your class.
3. Discuss: Should more space be made available for graffiti sprayers in your city? Person 1 is a sprayer. Person 2 is the mayor. Present the role-play to the class.

   (N 2, p. 260)

Within the dimension of social sustainability, one exemplary exercise asks students to reflect and write creatively about what is politically important to them on a local level, as well as current political issues on a national level in their own country. The exercise provides a direct link to reality, giving ideas on how the students' products from the exercise can be used in their real life and possibly have an impact outside of the classroom (Example 3).

(3)

1. What political issues are currently affecting you in your everyday life?
   a. At school?
   b. In your free time?
   c. In your private life?
2. What are the current political issues in Norway? Search the internet and find five headlines that deal with politics.
3.
   a. Write a Twitter note in which you briefly explain a current political issue. Your school probably has a Twitter account.
   b. Write a short text complaining about waste separation, too much plastic, the high prices in the canteen or on public transport, the long day at school, clothes that are produced too cheaply, etc. Your text is a post in a Facebook group.
   c. Write a short positive text in the school newspaper about the new sports facilities, the newly painted classrooms, the reasonably priced food in the school canteen, etc.

(N 2, p. 242)

In one of the exercises coded into the third subcategory, *conserving nature and wildlife* of ecological sustainability, students are asked to organise a fictional event for the environment and write an invitation to the event. Students are provided with useful phrases in the target language but are otherwise free to plan the event the way they see fit.

An exercise about favouring ethically sustainable products within the dimension of financial sustainability encourages students to think and discuss what clothes cost and what they should cost, taking into account the circumstances of their production (Example 4).

(4) Discuss the following in small groups:

1. What do you think fashion (pants, jackets, shirts, shoes, etc.) should cost? Should you have to spend a lot or a little on good brand name clothes?
2. Many people buy cheap fashion that is often produced by child labour in the Third World. What are your thoughts on this?
3. Many brands are fake copies of real fashion brands. What do you think about this?

(N 2, p. 167)

This example directs students to discover links between different dimensions of sustainability and to consider how our actions are linked to global issues. Reflecting on the questions could lead students to discuss and understand cultural, ecological, financial, and social aspects of consumerism. Many other exercises and texts include various subthemes of sustainable development, even from different dimensions, thus implicitly showing that sustainability needs to be thought of as a multidimensional concept.

Discussion

One finding that stands out from our results is that the coursebooks we analysed for this study covered a broad variety of sustainability topics. Five out of eleven books included all four sustainability dimensions, and three more included three out of the four dimensions. Most books focused more on the social and cultural dimensions, which is typical for language teaching (Maijala et al. 2023). Within all dimensions, we were able to find texts and exercises with an explicit focus on sustainability (Curdt-Christiansen 2021; Risager 2021) and which promoted a holistic understanding of sustainability as well as an action-oriented approach beyond the textbook and/or the learning situation. The results of the study indicate that more advanced level coursebooks have more SDG-related texts and activities than other levels of language coursebooks.

Exercises and texts belonging to the *advanced subcategory*, according to our coding structure, could be found throughout all four sustainability dimensions. In general, the analysed German coursebooks included a fair amount of SDG-related content (Lasekan et al. 2023). However, Agenda 2030 and SDGs were not explicitly mentioned in the coursebooks. The SDG-related content in German coursebooks mostly had an indirect connection to sustainability.

Exercises promoting a holistic understanding of sustainability encouraged students to reflect and express their opinions or take up someone else's perspective to discuss the subject at hand. These exercises included creative writing, which provided an opportunity for students to conduct research to obtain a better understanding of the context for the exercise. The coursebooks also regularly contained texts on the subject of ecological sustainability, showing good examples of treating nature and resources responsibly and exercises prompting independent thinking and being active. Through these activities, learners were encouraged to participate in environmental protection and transformational practices (Jacobs and Goatly 2000; Xiong 2014; Cristovão et al. 2022).

The German coursebooks focused, through self-descriptions of young figures, on the individual level; that is, they centred on what a young person could do in their everyday life (Risager 2021), favouring sustainable products and one's own consumption behaviour (Curdt-Christiansen 2021). In particular, the exercises related to the social dimension of sustainability, which encouraged learners to reflect on their actions and build connections between different perspectives (Sinakou et al. 2019). Texts and exercises within the topic of *preserving culture in all its forms* were coded into various subcategories. The texts and exercises that were not coded into the *advanced subcategory* within this topic mainly transported regional and cultural knowledge, for example, about traditions or food, not always explicitly pointing out cultural diversity.

One aspect mentioned frequently in the texts and exercises was volunteering, which was coded into *societal justice and well-being*. This happened, for instance, by giving examples of protagonists (Risager 2021) or other textbook characters. They performed or experienced exemplary actions that demonstrated how to engage in volunteer work. However, when giving different perspectives on volunteering, some coursebooks overgeneralised their target audience of learners in the Global North as providers and the Global South as receivers of volunteering activities. With this approach, textbooks can reproduce common but often harmful stereotypes (see Gray 2010; Risager 2021).

As coursebooks are the most commonly used source in language teaching (Guerrettaz and Johnston 2013), it is important to pay significant attention to their contents. The level of attention given to SDG-related content and ESD in textbooks varies considerably in different educational contexts and at different language levels (e.g., Mohammadnia and Moghadam 2019). Based on our findings, we can conclude that textbooks have great potential to promote ESD. Texts can inspire learners of all ages to learn about sustainability and reflect on their own actions as well as society's ways in general. According to our results, exercises that are beneficial for integrating ESD into language teaching encourage systems thinking by linking different dimensions of sustainability to one another and allowing learners to connect these dimensions directly to their own experiences and everyday lives.

As less research has been done on textbooks in languages other than English, more research is needed on their contribution to the promotion of ESD. Textbooks carry the responsibility to show the world not only through a middle-class perspective (e.g., Gray 2010), but also from the perspectives of different social backgrounds. One of the challenges that language teachers and textbook developers face is how to integrate SDG-related content and learner-centred reflective activities that promote ESD at the beginner's language level. This could be achieved, for instance, by using learners' first language in their reflections. Teachers' guides could also provide advice for language teachers on how to adapt and supplement textbook exercises to promote learner-centredness and reflection (Micalay-Hurtado and Poole 2022; Stibbe 2021). Furthermore, textbooks should contain more analytical and discovery exercises that enhance learners' linguistic and cultural experiences (Cristóvão et al. 2022; Risager 2022), which have been found to be beneficial in promoting ESD. One potential approach to enhancing the representation of SDG-related content in language textbooks is to integrate SDGs into existing topics or to create dedicated textbook units focused on them. Another strategy could be to explicitly inform readers that the objective of a specific text or exercise is to gain deeper insights into sustainability. In conclusion, it is essential that language teachers and language textbook developers, also in mainstream commercially produced coursebooks, address SDG-related issues in a more comprehensive manner (Risager 2021; Lewis et al. 2014; Sinakou et al. 2019). Doing so necessitates additional training for teachers and requires interdisciplinary collaboration in schools and publishing houses.

Appendix 1

Information and descriptions of the German language coursebooks used in the analysis

Title of the book (abbreviations in brackets)	Authors	Year of publication	Publisher	Size	Subdivision	Country of use	CEFR level
Auf uns. Grundbog til tysk i gymnasiet (DK 1)	Kate Fuchs, Mathilde Sinding	2020 (3rd edition)	Gyldendal	Coursebook (texts and exercises, 303 pages)	7 units	Denmark	B1–B2
Los geht's! (DK 2)	Mette Hermann	2020	Gyldendal	Coursebook (texts and exercises, 170 pages)	4 units	Denmark	B1–B2
Magazin.de 1 (FIN 1)	Pia-Helena Bär, Ines Paul, Ritva Tolvanen, Heidi Äijälä	2015	Otava	Coursebook (texts and exercises, 163 pages)	5 units	Finland	A1
Magazin.de 2 (FIN 2)	Pia-Helena Bär, Ines Paul, Ritva Tolvanen, Heidi Östring	2014	Otava	Coursebook (texts and exercises, 180 pages)	5 units	Finland	A1–A2
AQA GCSE German (GB 1)	Heather Murphy, David Riddel, Alan O'Brian, Helen Kent	2016	Oxford	Coursebook (texts and exercises, 207 pages)	12 units	Great Britain	Not specified

(continued)

Title of the book (abbreviations in brackets)	Authors	Year of publication	Publisher	Size	Subdivision	Country of use	CEFR level
Stimmt! Edexcel GCSE (9–1) German (GB 2)	Harriet Lanzer, Michael Spencer, Caroly Batstone, Lisa Probert	2016	Pearson	Coursebook (texts and exercises, 224 pages)	8 units	Great Britain	Not specified
Na klar! Duis voor de onderbouw 3 (NL 1)	Elisabeth Lehrner-te Lindert, Gabrie Gelissen, Danique Simons-Trommelen, Marit Molenaar	2020	Malmberg	Coursebook (texts and exercises, 176 pages)	6 units	The Nether-lands	Not specified
Neue Kontakte, Deutschbuch A (NL 2a)	Gordana Bozarov, Andrea Hahnfeld, Franziska Hünig, Clarije van IJssel-Groffel, Chris Linek, Sven Tillema, Jessica Verkerk-Schneider	2021	Noodhoff	Coursebook (texts and exercises, 164 pages)	8 units	The Nether-lands	A1–A2
Neue Kontakte, Deutschbuch A (NL 2b)	Maja Bakker, Gordana Bozarov, Andrea Hahnfeld, Franziska Hünig, Clarije van IJssel-Groffel, Chris Linek, Jessica Verkerk-Schneider	2021	Noodhoff	Coursebook (texts and exercises, 166 pages)	8 units	The Nether-lands	A2–B1
Klasse! 2 (N1)	Mona Gundersen-Røvik, Simen Braaten	2021	Cappelen Damm	Coursebook (texts and exercises, 167 pages)	10 units	Norway	Not specified

(continued)

(continued)

Title of the book (abbreviations in brackets)	Authors	Year of publication	Publisher	Size	Subdivision	Country of use	CEFR level
Weiter geht's (N2)	Anne-Marie Schulze	2020	Fagbokforlaget	Coursebook (texts and exercises, 384 pages)	15 units	Norway	Not specified

References

Auswärtiges Amt. 2020. *Deutsch als Fremdsprache weltweit. Datenerhebung 2020*. Available from: https://www.auswaertiges-amt.de/blob/2344738/b2a4e47fdb9e8e2739bab2565f8fe7c2/deutsch-als-fremdsprache-data.pdf Accessed July 2, 2024.

Biström, E., and R. Lundström. 2021. Textbooks and action competence for sustainable development: An analysis of Swedish lower secondary level textbooks in geography and biology. *Environmental Education Research* 27, no. 2 (January 2): 279–294. https://doi.org/10.1080/13504622.2020.1853063.

Borg, S. 2006. The distinctive characteristics of foreign language teachers. *Language Teaching Research* 10, no. 1 (January): 3–31. https://doi.org/10.1191/1362168806lr182oa.

Byram, M. 2022. Politics, origins and futures of the CEFR. *The Language Learning Journal* 50, no. 5 (September 3): 586–599. https://doi.org/10.1080/09571736.2020.1845392.

Carabantes, L., and A. Paran. 2022. 'I preferred to take another activity from the textbook': An activity–theoretical study of learning to design language teaching materials. *The Modern Language Journal* 106, no. 4 (December): 659–674. https://doi.org/10.1111/modl.12803.

Cohen, L., L. Manion, and K. Morrison. 2018. *Research methods in education*, 8th ed. New York, London: Routledge.

Corbin, J., and A. Strauss. 2014. Basics of qualitative research: Techniques and procedures for developing grounded theory. 4th Edition. California, London, New Delhi, and UK: Sage publications.

Council of Europe. 2001. *Common European framework of reference for languages: Learning, teaching, assessment*. Strasbourg: Council of Europe. Available from: https://rm.coe.int/1680459f97. Accessed October 6, 2024.

Council of Europe. 2020. *Common European framework of reference for languages: Learning, teaching, assessment—Companion volume*. Council of Europe Publishing, Strasbourg. Available from: www.coe.int/lang-cefr

Cristovão, V.L.L., B. Sanches, and G. Smart. 2022. Environmental discourse in Brazilian English-as-a-foreign-language textbooks: Socio-discursive practices and their implications for developing students' critical environmental literacy. *Environmental Education Research* 28, no. 1 (January 2): 75–94. https://doi.org/10.1080/13504622.2021.2007855.

Curdt-Christiansen, X.L. 2021. Environmental literacy: Raising awareness through Chinese primary education textbooks. *Language, Culture and Curriculum* 34, no. 2 (April 3): 147–162. https://doi.org/10.1080/07908318.2020.1797078.

EDUFI. 2019. *Lukion opetussuunnitelman perusteet 2019*. Available from: https://www.oph.fi/fi/koulutus-ja-tutkinnot/lukion-opetussuunnitelmien-perusteet. Accessed June 26, 2024.

Eilam, E., and T. Trop. 2010. ESD pedagogy: A guide for the perplexed. *The Journal of Environmental Education* 42, no. 1 (September 17): 43–64. https://doi.org/10.1080/00958961003674665.

Frisk, E., and K.L. Larson. 2011. Educating for sustainability: Competencies & practices for transformative action. *Journal of Sustainability Education* 2, no. 1 (March): 1–20.

de la Fuente, M.J., (ed). 2021. *Education for sustainable development in foreign language learning: Content-based instruction in college-level curricula*, 1st ed. New York: Routledge.

Gericke, N., B.J. Hudson, C. Olin-Scheller, and M. Stolare. 2018. Powerful knowledge, transformations and the need for empirical studies across school subjects. *London Review of Education* (November 15). https://doi.org/10.18546/LRE.16.3.06.

Gericke, N., L. Huang, M.C. Knippels, A. Christodoulou, F. Van Dam, and S. Gasparovic. 2020. Environmental citizenship in secondary formal education: The importance of curriculum and subject teachers. In *Conceptualizing environmental citizenship for 21st Century education,* eds. A.Ch. Hadjichambis, P. Reis, D. Paraskeva-Hadjichambi, J. Činčera, J. Boeve-de Pauw, N. Gericke, and M.C. Knippels, 4:193–212. Environmental Discourses in Science Education. Cham: Springer International Publishing. https://doi.org/10.1007/978-3-030-20249-1_13.

Gray, J. 2010. *Describing and analyzing ELT coursebooks. In The construction of English*. United Kingdom: Palgrave Macmillan UK.

Guerrettaz, A.M., and B. Johnston. 2013. Materials in the classroom ecology. *The Modern Language Journal* 97, no. 3 (September): 779–796. https://doi.org/10.1111/j.1540-4781.2013.12027.x.

Hahl, K. 2020. Kulttuurin muuttunut olemus opetussuunnitelmissa ja vieraiden kielten oppikirjoissa. [The transformation of culture in curricula and foreign language textbooks.] In *Kielididaktiikan katse tulevaisuuteen: Haasteita, mahdollisuuksia ja uusia avauksia kielten opetukseen,* eds. R. Hilden, and K. Hahl, 173–201. Helsinki: University of Helsinki.

Hahl, K., R.J. Longfor, and P.M. Niemi, (eds). 2015. *Diversities and interculturality in textbooks: Finland as an example*. Newcastle upon Tyne: Cambridge Scholars Publishing.

Hammadou, J., and E.B. Bernhardt. 1987. On being and becoming a foreign language teacher. *Theory Into Practice* 26, no. 4 (September): 301–306. https://doi.org/10.1080/00405848709543290.

Harjanne, P., C.D. Larenas, and S. Tella. 2017. Foreign-language teaching and studying in Chilean and Finnish classrooms as seen by teachers. *Journal of Language and Cultural Education* 5, no. 3 (September 1): 1–21. https://doi.org/10.1515/jolace-2017-0025.

Harwood, N. 2010. Issues in materials development and design. In *English language teaching materials: Theory and practice*, ed. N. Harwood, 3–30. Cambridge: Cambridge University Press.

Harwood, N. 2013. Content, consumption, and production: Three levels of textbook research. *In English language teaching textbooks: Content, consumption, production*. ed. N. Harwood, 1–41. United Kingdom: Palgrave Macmillan UK.

Jacobs, G.M., and A. Goatly. 2000. The treatment of ecological issues in ELT coursebooks. *ELT Journal* 54, no. 3 (July): 256–264. https://doi.org/10.1093/elt/54.3.256.

Jodoin, J.J., and J. Singer. 2019. A framework for integrating education for sustainable development in the English as a foreign language classroom in Japan: An appeal to the language teaching community. *Osaka JALT Journal* 6, no. 1 (June): 51–66.

Katunich, J., and J. Goulah. 2020. Introduction: TESOL and sustainability. In *TESOL and sustainability: English language teaching in the Anthropocene era*, 1st ed., ed. J. Katunich and J. Goulah, 1–16. London: Bloomsbury Publishing.

Lasekan, O.A., C.M. Méndez-Alarcón, B.S. Mathew, and E.S. Campos. 2023. Exploring the potential of a popular EFL textbook to foster both sustainability awareness and competencies among ESD learners: A content analysis approach. *Sustainability* 15, no. 16 (August 21): 12640. https://doi.org/10.3390/su151612640.

Lee, H.Y. 2014a. Inquiry-based teaching in second and foreign language pedagogy. *Journal of Language Teaching and Research* 5, no. 6 (November 1): 1236–1244. https://doi.org/10.4304/jltr.5.6.1236-1244.

Lee, J.F.K. 2014b. A hidden curriculum in Japanese EFL textbooks: Gender representation. *Linguistics and Education* 27 (September): 39–53. https://doi.org/10.1016/j.linged.2014.07.002.

Lewis, E., C. Mansfield, and C. Baudains. 2014. Ten tonne plan: Education for sustainability from a whole systems thinking perspective. *Applied Environmental Education & Communication* 13, no. 2 (April 3): 128–141. https://doi.org/10.1080/1533015X.2014.950890.

Littlejohn, A. 2022. The analysis and evaluation of language teaching materials. In *The Routledge handbook of materials development for language teaching*, ed. J. Norton and H. Buchanan, 263–276. London: Routledge.

Maijala, M., and M. Tammenga-Helmantel. 2019. Grammar exercises in Dutch, Finnish and global textbooks for teaching German as a foreign language. *The Language Learning Journal* 47, no. 5 (October 20): 537–557. https://doi.org/10.1080/09571736.2017.1309449.

Maijala, M., L.M. Heikkola, S.-R. Kuusalu, P. Laine, M. Mutta, and K. Mäntylä. 2023. Pre-Service language teachers' perceptions of sustainability and its implementation in language teaching. *Language Teaching Research* (May 13): 136216882311706. https://doi.org/10.1177/13621688231170682.

Maijala, M., N. Gericke, S.-R. Kuusalu, L.M. Heikkola, M. Mutta, K. Mäntylä, and J. Rose. 2024. Conceptualising transformative language teaching for sustainability and why it is needed. *Environmental Education Research*: 1–20. https://doi.org/10.1080/13504622.2023.2167941

McGrath, I. 2006. Teachers' and learners' images for coursebooks. *ELT Journal* 60, no. 2 (April 1): 171–180. https://doi.org/10.1093/elt/cci104.

Micalay-Hurtado, M.A., and R. Poole. 2022. Eco-critical language awareness for English language teaching (ELT): Promoting justice, wellbeing, and sustainability in the classroom. *Journal of World Languages* 8, no. 2 (October 3): 371–390. https://doi.org/10.1515/jwl-2022-0023.

Mochizuki, Y. 2019. Rethinking schooling for the 21st Century: UNESCO-MGIEP's contribution to SDG 4.7. *Sustainability: The Journal of Record* 12, no. 2 (April): 88–92. https://doi.org/10.1089/sus.2019.29160.

Mohammadnia, Z., and F.D. Moghadam. 2019. Textbooks as resources for education for sustainable development: A content analysis. *Journal of Teacher Education for Sustainability* 21, no. 1 (June 1): 103–114. https://doi.org/10.2478/jtes-2019-0008.

Öhman, J., and L. Sund. 2021. A didactic model of sustainability commitment. *Sustainability* 13, no. 6 (March 11): 3083. https://doi.org/10.3390/su13063083.

Patton, M.Q. 2015. *Qualitative research & evaluation methods: Integrating theory and practice*, 4th ed. Thousand Oaks, CA: SAGE.

Rathert, S., and N. Cabaroğlu. 2021. Teachers as slaves or masters to their coursebooks: An in-depth study on two English language teachers' coursebook utilization. *Language Teaching Research* 28, no. 5 (September): 1816–1841. https://doi.org/10.1177/13621688211036239.

Redman, A., and A. Wiek. 2021. Competencies for advancing transformations towards sustainability. *Frontiers in Education* 6 (November 30): 785163. https://doi.org/10.3389/feduc.2021.785163.

Richards, J.C. 2014. The ELT textbook. In *International perspectives on materials in ELT*, ed. S. Garton and K. Graves, 19–36. Basingstoke, UK: Palgrave Macmillan.

Riess, W., M. Martin, C. Mischo, H.G. Kotthoff, and E.M. Waltner. 2022. How can Education for Sustainable Development (ESD) be effectively implemented in teaching and learning? An analysis of educational science recommendations of methods and procedures to promote ESD goals. *Sustainability* 14, no. 7 (March 22): 3708. https://doi.org/10.3390/su14073708.

Risager, K. 2021. Language textbooks: Windows to the world. *Language, Culture and Curriculum* 34, no. 2 (April 3): 119–132. https://doi.org/10.1080/07908318.2020.1797767.

Risager, K. 2022. Culture and materials development. In *The Routledge handbook of materials development for language teaching*, ed. J. Norton and H. Buchanan, 109–122. London: Routledge.

Satokangas, H., and P. Mikander. 2023. From a welcoming rhetoric to a narrowing one: Constructing citizen agency in Finnish social studies textbooks. *Nordidactica: Journal of Humanities and Social Science Education* 13, no. 3 (June): 85–108.

Sinakou, E., V. Donche, J. Boeve-de Pauw, and P. Van Petegem. 2019. Designing powerful learning environments in Education for Sustainable Development: A conceptual framework. *Sustainability* 11, no. 21 (October 28): 5994. https://doi.org/10.3390/su11215994.

Stibbe, A. 2004. Environmental education across cultures: Beyond the discourse of shallow environmentalism. *Language and Intercultural Communication* 4, no. 4 (November 15): 242–260. https://doi.org/10.1080/14708470408668875.

Stibbe, A. 2021. *Ecolinguistics: Language, ecology and the stories we live by*. 2nd Edition. Abingdon, Oxon; New York, NY: Routledge.

Sund, P., N. Gericke, and G. Bladh. 2020. Educational content in cross-curricular ESE teaching and a model to discern teacher's teaching traditions.

Journal of Education for Sustainable Development 14, no. 1 (March): 78–97. https://doi.org/10.1177/0973408220930706.

The Earth Charter. 2000. *The Earth charter in action 2000 annual report*. Available from: http://earthcharter.org/virtual-library2/the-earth-charter-in-action-2000-annual-report/. Accessed August 6, 2024.

Tomlinson, B. 2012. Materials development for language learning and teaching. *Language Teaching* 45, no. 2 (April): 143–179. https://doi.org/10.1017/S0261444811000528.

Tomlinson, B. 2022. The discipline of materials development. In *The Routledge handbook of materials development for language teaching*, ed. J. Norton and H. Buchanan, 3–16. London: Routledge.

Tomlinson, B., and H. Masuhara. 2013. Adult coursebooks. *ELT Journal* 67, no. 2 (April 1): 233–249. https://doi.org/10.1093/elt/cct007.

Ulum, Ö.G., and D. Köksal. 2021. A critical inquiry on the ideological and hegemonic practices in EFL textbooks. *Multicultural Learning and Teaching* 16, no. 1 (March 31): 45–61. https://doi.org/10.1515/mlt-2019-0005.

UNESCO. 2018. *Global action programme on Education for Sustainable Development. Education for Sustainable Development and the SDGs. Policy brief: Advancing ESD policy.* Available from: https://unesdoc.unesco.org/ark:/48223/pf0000246270. Accessed July 22, 2024.

UNESCO. 2020. *Education for Sustainable Development: A roadmap.* France: UNESCO.

Vásquez, C., I. García-Alonso, M.J. Seckel, and Á. Alsina. 2021. Education for Sustainable Development in primary education textbooks——An educational approach from statistical and probabilistic literacy. *Sustainability* 13, no. 6 (March 12): 3115. https://doi.org/10.3390/su13063115.

Vu, M.T., and T.T.T. Pham. 2022. Gender, critical pedagogy, and textbooks: Understanding teachers' (lack of) mediation of the hidden curriculum in the EFL classroom. *Language Teaching Research* (November 14): 13621688221136937. https://doi.org/10.1177/13621688221136937.

Weninger, C., and T. Kiss. 2013. Culture in English as a foreign language (EFL) textbooks: A semiotic approach. *TESOL Quarterly* 47, no. 4 (December): 694–716. https://doi.org/10.1002/tesq.87.

Wiek, A., M. Bernstein, R. Foley, M. Cohen, N. Forrest, C. Kuzdas, B. Kay, and L. Withycombe Keeler. 2015. Operationalising competencies in higher Education for Sustainable Development. In *Handbook of higher education for sustainable development*, ed. M. Barth, G. Michelsen, M. Rieckmann, and I. Thomas, 241–260. London: Routledge.

Xiong, T. 2014. Shallow environmentalism: A preliminary eco-critical discourse analysis of secondary school English as a Foreign Language (EFL) texts in China. *The Journal of Environmental Education* 45, no. 4 (October 2): 232–242. https://doi.org/10.1080/00958964.2014.943686.
Zahoor, M., and F. Janjua. 2020. Green contents in English language textbooks in Pakistan: An ecolinguistic and ecopedagogical appraisal. *British Educational Research Journal* 46, no. 2 (April): 321–338. https://doi.org/10.1002/berj.3579.

7

Transformative Language Teaching Practices for Sustainability Among Adult L2 Learners of Swedish in a German University: The Case Study as a Teaching Method

Marianne Broermann, Salla-Riikka Kuusalu, and Minna Maijala

Introduction

The teaching of sustainability and Education for Sustainable Development (ESD) in higher education institutions (HEIs) began to take shape primarily in the late 1980s and early 1990s, although the importance of education in environmental matters had been highlighted in the Declaration and Action Plan of the UN Conference on the Human Environment in Stockholm in 1972 (United Nations 1973). Environmental education began taking root largely as a response to environmental concerns and a resulting growing global environmental awareness. This period saw the emergence of environmental studies programmes and the

M. Broermann
Department of Scandinavian Studies, University of Göttingen, Göttingen, Germany

Centre for Languages and Transferable Skills, University of Göttingen, Göttingen, Germany

S.-R. Kuusalu (✉) · M. Maijala
School of Languages and Translation Studies, University of Turku, Turku,

M. Maijala et al. (eds.), *Transformative Language Teaching for Sustainability*, https://doi.org/10.1007/978-3-031-85493-4_7

integration of sustainability into various academic disciplines. However, the focus lay on environmental aspects, although the Brundtland Report (WCED 1987) saw sustainability as consisting of three dimensions: economic, social and environmental. Since the adoption of Agenda 2030 and its 17 Sustainable Development Goals (SDGs) in 2015 (United Nations 2015), the notion of 'sustainability' has become widespread, and the efforts of the education system to take action towards reaching the SDGs have become more comprehensive in teaching, research and community engagement. HEIs play a central role in the transformation towards a more sustainable society, both in generating and transferring relevant knowledge and educating future decision-makers (Barth 2014; Leal Filho et al. 2016).

A study by the German Environment Agency (Umweltbundesamt 2022) found that young people in Germany had clear attitudes towards climate protection and a high willingness to change. However, their environmental behaviour was below average. Vallabh (2018) also found a strong openness among young people to the need for major changes in the systems responsible for sustainability problems. Furthermore, empirical research has demonstrated a clear desire for ESD by young people (Michelsen et al. 2016; Berglund et al. 2014; Boewe-de Pauw et al. 2015; Grund and Brock 2020). However, the existing research has focused on the environmental component of sustainability and paid less attention to economic aspects, not to mention the social and cultural aspects, which are the focus in language education (Maijala et al. 2024). In this study, the case study method was used as a pedagogical framework to allow the issue to be approached from different angles, including from socio-political and cultural perspectives.

This chapter presents a teaching project from a university language course on Swedish in Germany in which ESD-related content and methods were used. The aim of the study was to determine whether and

Finland
e-mail: srkbos@utu.fi

S.-R. Kuusalu
Department of Biology, University of Turku, Turku, Finland

in what way the language course had altered the students' conceptions of sustainability. First, the role of ESD in HE in general and language teaching in particular are discussed. Second, the case study as an ESD language teaching method and the use of reflective tools in language learning are considered. This is followed by a presentation of the study: its context, methodology and findings. Finally, the results of the study and the teaching approach used are presented.

Education for Sustainable Development in Higher Education

As Bianchi (2020) has stated, a great number of terms are used, often interchangeably, in the literature to refer to sustainability in education, for example, *Education for Sustainable Development*, *education for sustainability* and *sustainability education*. In this article, the term ESD is used to highlight the role of education in reaching the SDGs, as 'education is a way to equip students with the necessary set of knowledge, skills, attitudes and values throughout their lives to enact a sustainable development (or progress or growth)' (Bianchi 2020, p. 11). In accordance with Bianchi (2020), the term *sustainability* is seen as a long-term goal and *sustainability education* as a pathway to reaching a more sustainable world, with ESD as a central part of it.

In her overview of the literature on teaching sustainability, Molina (2022) highlighted the edited volumes of Melin (2019) and de la Fuente (2021), both of which demonstrate methods for integrating ESD into language teaching. In 2022, Argondizzo and Mansfield (2022) edited a special issue on ESD, where a number of examples can be found on the integration of sustainability topics in language courses in HE. Further, Gabaudan (2022) reported on a restructured course module in French based on the method of reflective practice. She showed how a language course could be structured around SDGs, how it enhanced student engagement and language skills and how the course content, learning objectives, and assessment could be aligned. Hämäläinen (2022) showed how the use of the concept of the Bohm dialogue connected with reflection tasks in an English course could contribute to the ethical dialogue skills needed for sustainable development. In her article, Kaur (2022) showed how critical pedagogy could be used to accommodate learning that disrupts students' thinking and encourages political engagement. Marcella and Samofalova (2022) showed how data driven, and research-based approaches applied to a corpus, such as a collection of Instagram posts, could contribute not only to climate change awareness but also research and language skills among students. These examples show that the consideration of ESD in language teaching and learning is gaining momentum. There are various reasons why language teachers have been reluctant to teach ESD, including insecurity about their own ESD competence and lack of time or teaching materials (Maijala et al. 2024). De la Fuente (2021) lists relatable challenges in US college-level language departments: the disconnection between language and literature, language and content, lower-level versus higher-level courses, etc.

Interdisciplinary teaching of ESD provides a variety of perspectives, and SDGs provide a wide range of topics for language teaching. In addition, language learning situations provide a natural and realistic setting for training intercultural skills, co-operation, ethical communication and other sustainability-related competencies.

The Case Study as a Language Teaching Method in Education for Sustainable Development

Addressing current and future sustainability challenges requires individuals who are able to collaborate, manage complexity, tolerate ambiguity and handle urgent problems that lack an obvious optimal solution (Sprain and Timpson 2012). It has been suggested that interdisciplinary, holistic, multi-method and values-driven pedagogical approaches promote alternative, student-centred methods that actively engage students, encourage critical thinking and foster reflection (UNESCO 2006). The case study as a pedagogical approach fits this description especially well, as argued by Lozano et al. (2017, p. 7):

> In case studies, qualitatively rich descriptions of settings, problems, and controversies in sustainable development challenge students to interact with the inherent complexity and uncertainty found in global, regional, and/or local contexts. Case studies invite students to consider real-world examples and examine issues from a diversity of stakeholder perspectives.

Based on a literature review of publications on sustainability education and pedagogical approaches, Lozano et al. (2017) constructed a framework to illustrate the connections between different pedagogical approaches and sustainability competencies. The resulting framework showed that case studies and problem-based learning as universal pedagogical approaches, especially when used together, have the strongest potential to cover most of the sustainability competencies, such as critical thinking and systems thinking. In contrast, lecturing (as a more 'traditional' approach) alone does not seem to promote sustainability competencies, although it may be included as a component of other approaches (Lozano et al. 2017). The role of systems thinking as a pedagogical approach is also highlighted by the World Economic Forum (WEF 2023) in a white paper on innovative learning solutions to educate citizens to tackle complex global challenges. It therefore advocates the mainstreaming of systems thinking as a framework at all levels of education and provides a number of practical examples of how future education could adopt a systems thinking perspective.

When a case study presents a challenge or problem to be solved, it can also be classified as a problem-based learning arrangement (see Fischer 2008; Lyu 2023). A problem-based case study aimed at training students' sustainability competencies reflects a systems approach in which the components of sustainability (ecological, economic, social, cultural) are dealt with. In such case studies, students are invited to investigate a problem from multiple perspectives, many of which are open-ended and do not have a prescribed outcome—only a complex problem (see Sprain and Timpson 2012; Lozano et al. 2017). Case studies have long been utilised in language learning, particularly in languages for specific purposes (LSP), such as business English (Lyu 2023). Fischer and others (2008, 2009) outlined the process of a language case study in different

HEI language course contexts, emphasising a holistic approach that integrates various language skills and promotes teamwork and problem-solving. They highlighted that the authenticity of case studies enhances student engagement, interest, and learning success (Fischer et al. 2008). In contrast to random listening and reading exercises, targeted exercises have felt more meaningful to students (Fischer et al. 2009). This approach is also in line with the task-based approach of the Common European Framework of Reference for Languages (Council of Europe 2001) and its Companion Volume (Council of Europe 2020).

As a pedagogical approach, language case studies are similar to case studies that promote sustainability competencies. The sustainability skills being trained are more or less the same, although the subject knowledge level may differ, as students in general language courses are mixed and represent a multiplicity of disciplines. Another difference is that in a language case study, the focus is on language and communication (see Boyd 1991). Thus, the specific objectives of a language case study with a sustainability topic would include contributing to students' knowledge of the case study topic to help them understand the complexity of sustainability-related problems and enable them to communicate, discuss and help solve such problems through the medium of the target language.

The Use of Reflective Tools in Language Learning

Reflective practice has been shown to have a positive impact on students' performance in university language classes by strengthening their role and engagement in the learning process (Torres-Goens and Farley 2017). Reflective tools, which include diaries, videos, portfolios, and blogs, foster constructivist learning environments and encourage learners to adopt a more curious and open approach to learning (Tuncer and Özkan 2021). Implementing reflective approaches in HE language learning also benefits students by increasing their awareness of their learning and building on their learning experiences.

This type of reflection is called reflection-on-action (Schön 1987), that is, thinking back after having engaged in a task, project, etc. As the act of reflection requires the retrieval, elaboration and generation of information, it can make learning more sustainable for students and foster critical thinking skills (Brookfield 2012). At one end of the non-reflection–reflection continuum of Peltier and others (2005), non-reflection learning refers to a surface approach to learning characterised by the quick solving of tasks without attaching personal or cognitive meaning to the tasks, relying on memorisation and generally focusing on what is required rather than exploring additional learning materials. In contrast, reflection learning is deep learning; it shows active and conscious engagement, identification with the tasks and careful consideration of new information. Brown and others (2014) presented an example of a reflection practice that is easy to adapt to different learning contexts. Here, very general short prompts, such as 'What went well? What could have gone better? What other knowledge or experience does this remind you of?' etc., are used to help learners initiate reflection.

Research Context and Methodology

This study was conducted at a German university language centre in two language courses with students learning Swedish at the CEFR level B1/B1+ in the summer terms of 2023 and 2024. There was a great deal of variety among the students' main study subject: humanities (languages and literatures, theology, digital humanities), social sciences (anthropology, education, political science, law), economics and natural sciences (forestry, biology, biodiversity, ecology). As mentioned earlier, this can be regarded as a strength in terms of multiple viewpoints regarding the task presented in the case study.

The study aimed to answer the following research questions:

1. Did the language course change the students' conceptions of sustainability; if so, in what way?
2. To what extent could deeper reflection and transformative learning be identified as a result of the course project?

Both student questionnaires and end-of-course reflective writings were used as study data. The questionnaires were presented using an electronic form to the students before (Questionnaire 1, Q1) and after (Questionnaire 2, Q2) the case study task in German, which was known to be the L1 of all course participants (see Appendix 1 and 2 for Q1 and Q2, respectively; see also the questionnaire in Chapter 8).

A reflective writing task (Appendix 3) was used at the end of the course for two purposes: 1) to assess students' language learning outcomes and 2) allow students to reflect on their learning experience regarding the sustainability topics covered in the language course. In their writings, the students answered questions in Swedish related to the case study task content, their language learning, collaboration with each other and their personal learning. This home assignment contained similar questions to those proposed by Brown and others (2014): what they had learnt in terms of language and topic content, what had gone well/not so well and why and what insights they had. In Table 7.1, we see that most students responded to both questionnaires and submitted a reflective text at the end of the project. As there were a total of 15 complete responses to both the questionnaires and reflective writings, the basis for analysis was set at 15 participants.

The participants consented to the use of their responses for research purposes, and all examples were translated from German into English by the authors. Responses to Q1 and Q2 for the years 2023 and 2024 were analysed anonymously: the identification label consists of the identification number, year and the questionnaire number (e.g., ID03–23–Q2). During the data collection and analysis, the reflective writings were not treated anonymously, but they were later anonymised. We refer to the

Table 7.1 The number of submitted reflective writings and responses to questionnaires 1 (Q1) and 2 (Q2) in the courses held in 2023 and 2024

	2023 (n = 7)	2024 (n = 11)	Total (N = 18)
Q1	7	10	17
Q2	7	8	15
Reflective writings	6	9	15

students' reflective writings as follows: e.g., in ID04–24, ID04 is the student identification number, and 24 means reflective texts from 2024.

The two 14-week language courses covered a number of topics, including many sustainability issues, such as diversity and inclusion, national minorities and minority languages in Sweden, environmental and climate issues, and consumption. The student tasks and the teaching methods employed in the courses varied throughout the course, always including both productive and receptive activities, mediation, discussion and reflection. This case study was used as a teaching method during the topic 'Grön omställning i Sápmi' (the green transition in Sápmi, see European Commission 2020; Sámiskt informationscentrum 2024). Sápmi means the cultural regions that are traditionally inhabited by the Sámi people in the Northern parts of Fennoscandia. The period when the case study method was used covered three weeks (6 × 90 min instruction in class) of the language course, and an approximation of the teaching procedure described by Fischer and others (2008) was used (see Table 7.2). However, during the search for information, the students were confronted with authentic material online, such as media articles and reports, government reports, strategy papers of wind power and mining companies, etc. They were allowed and encouraged to use online tools to facilitate understanding when needed, such as online dictionaries and DeepL Translate.

Findings

Concept of Sustainable Development

For almost all students, the concept of sustainable development had an intergenerational dimension (see Example 1): sustainable development was characterised as development that does not jeopardise the livelihood of future generations, which is in line with the Brundtland Report (WCED 1987). Most students, however, focused their definitions on the ecological aspect of sustainability, such as natural resources (Example 2), and only a few considered social, economic or ideological/political aspects (Example 3).

Table 7.2 Descriptions of the case study task and data collection process during the course 'Grön omställning i Sápmi'

Week	Teaching/learning content		Data collection
1	Context definition and problem presentation	Context: Green transition in Sápmi, need for energy, electrification Input on Sápmi: colonisation and colonialism, reindeer husbandry as culture and language carrier of the Sámi (Sámediggi 2021) Overall problem to be solved: 'How can the increasing demand for renewable energy and strategic raw materials be met without jeopardising Indigenous peoples' rights to land, livelihood and culture, especially in Sápmi? Is the green transition also green colonialism?' (Lawrence 2014; Normann 2021; Össbo 2023; Drugge 2021; Nachet et al. 2022; Össbo 2023; Sköld 2011)	Q1

(continued)

Table 7.2 (continued)

Week	Teaching/learning content		Data collection
2	Forming of groups and start of group work	The students were divided into groups and asked to find information about the positions of different stakeholders The questions of the case study task were as follows: • What interests do the various stakeholders represent? • All stakeholders argue in some way that they are acting in a sustainable manner; what are their arguments? Which of the SDGs are related to the different stakeholder arguments? • What (sustainable) solutions do the different parties have to present in order to reach a compromise? How can these solutions be integrated? • What conclusions can be drawn from this and similar conflicts for current and future mineral, energy, etc., extraction in local/regional, national and global contexts?	

(continued)

Table 7.2 (continued)

Week	Teaching/learning content		Data collection
2–3	Students' research task	Materials from the teacher and students' own search for sources, preparation of reference lists (in and outside of class)	
3	Presentations & discussion	Group presentations on stakeholder positions. Roundtable discussion	Q2
after 3	Reflection & feedback	Reflective writing assignment on the sustainability topics covered during the course, feedback on the presentations and discussion (outside of class)	Reflective writing task

(1) Sustainable development includes the needs of future generations in decision-making processes. [...] (ID04–24–Q1)
(2) I understand sustainable development as a resource-conserving approach to the environment. This includes various things. A CO2-neutral climate policy is also part of it. [...] (ID06–23–Q1)
(3) I understand sustainable development as political and social developments that take social, and ecological but also cultural factors into account [...] (ID09–24–Q1)

One reason for the emphasis on environmental aspects may be that in both Swedish courses, the topics discussed before the case study project dealt with climate, consumption, waste and other environmental issues. This may have made the students more conscious of the environmental aspects of sustainable development than other aspects. This is also the aspect that most research has focused on (e.g., Michelsen et al. 2016; BMUW 2021; Boeve-de Pauw 2015).

When asked in Q2 whether their understanding of sustainable development had changed after the case study project, most students answered that it had not substantially changed but that they had learnt more about the issue, especially in relation to contradictions between sustainability goals. The answers show that the students became aware that there was no straightforward solution to the question in the particular case because of the complexity of the problem and the many legitimate interests of the various stakeholders (Example 4).

(4) My understanding of sustainable development has definitely changed [...] it has made me more thoughtful about the complexity of different factors that play a role in this overall constellation. I now better understand [...] that several legitimate interests can conflict with each other. (ID09–24–Q2)

Students' Expectations and Goals

When asked about their expectations with regard to knowledge on sustainability matters, the students mentioned a better understanding of complexity (Example 5) and language competence (Example 6). A majority of the students stated that their expectations and goals were met, both in terms of knowledge on sustainable development and the complexity of the issue and language competence. (Example 7).

(5) I expect to find out more about the different sides of the issue and to be able to form an opinion in the end, and perhaps even find a compromise. (ID04–24–Q1)
(6) Raise awareness through information & discussion, questioning decisions/developments, improve key vocabulary/discussion skills. (ID01–23–Q1)
(7) In the course section, I gained comprehensive insights into important parts of Swedish environmental policy from different perspectives, which is why my expectations of the teaching unit were fulfilled. I was also able to expand my Swedish vocabulary on the topic of sustainability, which was another important goal for me. (ID09–24–Q2)

Interest in the green transition

Both before and after the case study task, the students said that they were most interested in strategies to solve the actual and other sustainability-related problems and the different stakeholder positions (Example 8). Their answers show that they had been made aware that there was no one simple answer, and some were even able to reflect on the discourses about the topic (Example 9) or the worldview of the Sámi as an Indigenous people (Example 10).

(8) What interests me most are the ideas and approaches that have been/are being developed to solve the problem. By that I mean all the possibilities that are considered (even those that are not ultimately realised)—for the reason that you get to know so many possibilities, so many points of view and therefore have a better overview. (ID03–24–Q1)

(9) I found the different perspectives from which the dilemma was viewed incredibly exciting. In fact, I was able to understand all the perspectives, and everyone involved has good arguments. The only question is how they are weighed up against each other. I also found the question of discourse sovereignty fascinating. Who speaks? Who is heard? Who is not heard/overheard/ignored?? (ID07–24–Q2)

(10) I was particularly interested in the Sámi perspective on 'grön omställning', as their opinions and feelings should be given a more prominent place due to their status as the only Indigenous people in Europe. I was particularly fascinated to learn more about the Sámi's own views on the environment and nature as well as their visions of a sustainable future, which stand in stark contrast to the Western, profit-oriented economic model. (ID09–24–Q2)

Reflective Writings

All students adhered to the expected text length, resulting in texts of 280–360 words, and the structure suggested in the task description, starting with facts, followed by reflection on the sustainability topics and the learning outcomes. However, the level of reflection varied considerably, from a rather simple activity report with an evaluation to a deeper reflection on their own learning and thinking.

'Sápmi' was the most frequently mentioned issue in the students' reflective writings. This is not surprising, as this was the last topic related to sustainable development before the writing assignment. Table 7.3 shows that the sustainability topics examined during the course (Sápmi, energy, national minorities, consumption and diversity) were addressed in the students' reflective texts. Some of the topics overlapped more or less, for example, the Sámi are one of Sweden's five national minorities, a topic discussed earlier in the course. The energy issue was discussed as an introduction to the case study on the green transition in Sápmi and was also a central component of the case study. The environmental issues treated earlier in the course examined, in particular, consumption and waste.

In addition, Table 7.3 shows how often the students reported on their language learning, the possible increase in their knowledge, the pedagogical approaches used, whether there was any transfer between their subject study and the sustainability topics in the Swedish course and whether the study of the sustainability topics in the Swedish course had had an effect on their thinking. All 15 students reported that they had gained new knowledge, with the majority also stating that the complexity

Table 7.3 Topics addressed by the students (N=15) in their reflective texts

Topic	Number of mentions
Increase of knowledge	15
Language learning	13
Pedagogical approach	12
Sápmi	10
Consumption	7
Transfer	6
Personal development	5
Energy	4
National minorities	4
Diversity	2

of the issues had become clearer to them (Example 11), which was also identified as a key finding in the questionnaires.

(11) I learned a lot of interesting facts about the Swedish situation. I didn't know much about the Swedish government and how it acts. Also, I had no idea about the Sámi culture. I understand the conflict much better now but also find it quite complicated to find a solution—maybe impossible. (ID04–24)

Almost all students commented on the language work undertaken during the project. They also stated that they had learnt new vocabulary, grammar, etc., but some of them found the material linguistically challenging. This is understandable, as the students were at language learning level B1/B1+ and were confronted with authentic material. However, this challenge was also seen as positive (Example 12).

(12) Overall, it was sometimes a bit difficult to understand all the things we had to read and watch because it required many new words, but I still think it was good to at least try, even if you may sometimes not understand everything that is said. At the same time, it felt good that sometimes you could understand real documentaries or websites that were not in easy Swedish, so it motivated me a lot. (ID05–23)

The pedagogical approach and learning activities received positive comments from most students. They felt that the presentations and discussions improved their speaking skills (Example 13). Working independently on a collaborative project was also appreciated (Example 14).

(13) For me, the most important thing is to get better at speaking Swedish, and that's why I found presentations and discussion groups most useful. (ID02–23)
(14) While working on the topics, I liked that we had many tasks we could work on independently [...] because I think that through such tasks, you learn the most. Also, it was good for my confidence

in my Swedish skills to see that I could talk about the complicated topics in front of other people. (ID06–24)

As the students came from different fields of study, the connections between the subject study and the topics covered in the Swedish course may have been more obvious in some subjects than in others. Notwithstanding, six of the 16 students stated that there had been some transfer (Example 15).

(15) The writing task on climate or environmental issues was an easy topic for me because I could use the knowledge of my studies for my master's thesis and lectures a bit. (ID07–24)

According to Peltier and others (2005), the deepest level of learning, which includes intensive reflection, has the potential to transform our ways of thinking and personal beliefs, even action. 'Personal development' (a component of the reflection in the student texts) was demonstrated only five times (see Examples 16 and 17), which seems to be a fairly low share. This may be due to a number of factors, one being that the students were not explicitly asked to reflect on their learning in a systematic way, although various forms of feedback were collected and exchanged throughout the course. Furthermore, the extent to which reflective learning and teaching are practised in different subjects is likely to vary considerably.

(16) In addition, I now have a new perspective on Europe and why it is so difficult to be a community. Sápmi is just an example of why there is rarely right and wrong—the green transition is important for the climate, but not at the expense of the Sámi. So what can be done? There is no perfect solution. Now I have the opportunity to look at the situation in other countries. Although I have grown up and live in Germany, I know nothing about the concerns of minorities here. What do they think about life here? (ID01–24)
(17) In this way, I learned even more to reflect and deconstruct my personal thought patterns with regard to ethnocentrism and not

> necessarily take my own culture's worldview as universal and all-encompassing, including in the context of nature and climate protection. (ID05–24)

Discussion

As expected, the students showed a relatively high level of awareness of sustainability issues and an openness to and interest in learning more about the topic. However, the focus was often on the environmental aspect, with less attention paid to the economic and social aspects, which were highlighted in this case study method. The case study seems to have increased the students' knowledge of how sustainability issues are dealt with in Sweden as well as the specific case of Sápmi. Perhaps most importantly, the case study raised the students' awareness of the complexity of the issue, where clean energy production seems to threaten the survival of the culture and livelihood of an Indigenous people. It could even be argued that some SDGs seem to be in conflict with each other.

Teaching controversial issues can promote transformative learning and sustainable behaviour (Grund and Brock 2020). The case study method was used to enable students to consider different solutions and reflect on the situation both locally and within a global context. In the case of Sápmi, the students concluded that they could not find a solution but that this was not the point. Rather, they learned that there were a variety of legitimate positions, pointing to the highly complex nature of the issue. They may have sympathised or even identified with one or more of the stakeholders in the case, but they were asked to maintain objectivity, which is a guiding principle of academic research.

As stated earlier, a case study conducted in a language learning course does not usually reach the depth of a case study in a subject study course; the objective of a language case study is to increase students' knowledge of the case study topic and enable adequate communication to discuss (and eventually solve) such problems in the target language. In a case study, students read and listen to complete productive tasks in effective group work, such as giving presentations and participating in panel discussions. In such activities, they are engaged in mediating their knowledge to others. The students' answers on the questionnaire and their reflective writings supported this approach: they stated that they learned best by engaging in discussions and giving and listening to presentations, combined with individual language work. This kind of collaborative work seems to support both language and other (sustainability) competencies, such as problem-solving, collaboration, etc., and could easily be utilised in other educational contexts also outside Europe.

Some of the students found the material linguistically challenging, but this was not seen as entirely negative. Many of them also felt that the challenging material helped them improve their language skills. This shows the importance of the teacher's role as a facilitator in helping students with language problems. The level of the Swedish courses was 'only' B1/B1+ , and only part of the input materials was adapted to this level, especially those that were also used for vocabulary and grammar training. Here, the students practised a number of mediation and reading strategies, such as breaking down complicated information into smaller units and making it accessible to others, both in terms of language and content. Some of the materials they used were in English, which meant that they had to mediate between languages using their plurilingual competence (see descriptors for mediation and reception in Council of Europe 2020). In doing so, they could develop skills that were not limited to foreign language learning but also applicable in a wide range of academic and everyday contexts. Understanding linguistically challenging material could also be a collaborative activity.

Most students followed the prescribed structure in their reflective writing, first describing what they had done, then commenting on and evaluating the tasks, their group and individual work and the pedagogical approach used. Here, as in the questionnaires, they indicated that they were relatively familiar with sustainability issues but that they had gained new knowledge and, above all, an awareness of the complexity of the issue, thus demonstrating traces of systems thinking skills. This supports the appropriateness of the case study as a teaching method that addresses systems thinking skills (Lozano et al. 2017). However, only a few of the texts showed a deeper level of reflection, where students commented on their way of thinking or worldview (Peltier et al. 2005). This indicates that in the case study method, reflection is not a very successful means of learning. However, because deep reflection seems most likely to lead to deep learning and transformation (Brookfield 2012; Peltier et al 2005), training becomes necessary as well as the design of course concepts and the adaptation of pedagogical approaches.

According to Grund and Brock (2020), knowledge alone is not a very strong predictor of behaviour; instead, it is mediated by other determinants, such as emotions and values. This is why a case study as a language teaching method is particularly suitable for integrating language learning and SDG-related content and methods, as it enables students not only to increase their knowledge of the topic but also mediate, discuss and evaluate different positions and possible solutions to the problem in the target language both during and at the end of the task. Combined with reflective tasks, it can be a useful pedagogical approach to simulate many current and future communication situations, where many, often conflicting, views need to be reconciled.

Appendix 1

Questionnaire 1: Q1–23 and Q1–24 (translated from German to English by the authors); * indicates compulsory questions

1. Year of birth *
2. Mother tongue/s *
3. Main study subject/s *
4. What do you understand by sustainable development? *
5. What are your expectations and goals for the course section 'grön omställning?'
6. Which subject area of the course section 'grön omställning' interests you the most and why? *
7. Which of the working methods listed below do you think are most helpful with your language learning either in the classroom or outside? You can choose a maximum of three alternatives. *
    - Project work
    - Group work
    - Independent work
    - Pair work
    - Interactions with artificial intelligence
    - Other; please specify
8. Which of the exercises or tasks listed below do you think are most helpful with your language learning either in the classroom or outside? You can choose a maximum of three alternatives. *
    - Simulations and role-plays
    - Conversations in the language of learning, self-evaluation and reflection and creative tasks
    - Practical exercises and listening exercises
    - Reading exercises, vocabulary exercises and writing exercises
    - Practising with native speakers and language learning apps
    - Competitions and games
    - Creating and making presentations and grammar exercises
    - Other; please specify

9. In the field below, you can write further thoughts about the course, for example, suggestions for improvement and what you would like to see in future courses.
10. My answers may be used for research purposes. *

    - Yes
    - No

Appendix 2

Questionnaire 2: Q2–23 and Q2–24 (translated from German to English by the authors); * indicates compulsory questions

1. Year of birth *
2. Mother tongue/s *
3. Main study subject/s *
4. Has your understanding of sustainable development changed since the course section 'grön omställning?' If so, please tell us how. *
5. What expectations were met, and what goals did you achieve during the 'grön omställning' section of the course?
6. Which of the topics in the course section 'grön omställning' were you particularly interested in and why? *
7. Which working methods used in the course section 'grön omställning' do you think best enabled you to learn the language? List 1–3.
8. Which of the tasks or exercises in the course section 'grön omställning' do you think were most useful for your language learning? List 1–3.
9. In the field below, you can write further thoughts about the course, for example, suggestions for improvement and what you would like to see in future courses.
10. My answers may be used for research purposes. *

    - Yes
    - No

Appendix 3

Reflective writing task (translated from Swedish to English by the authors); * indicates compulsory questions

Reflect on the sustainability issues that we have addressed, including the following:

- Diversity and inclusion
- National minorities and minority languages
- Environmental issues and sustainability
- The green transition in Sápmi.

You can either write more generally about several of these themes or simply select one thematic area to focus on. The following questions can be helpful:

- What was the topic about? Provide a brief description/overview.
- What was new to you? What was not?
- What did you learn in terms of content and language? Did you learn anything about yourself or others? What would you have liked to learn more about, or would you have preferred alternatives to what we did?
- How would you describe the tasks delegated to you? Was there co-operation? What did you like/dislike about the tasks and/or the level of co-operation? Why was this the case?
- What conclusions can you draw from what you have worked on? Did you gain new insights?

References

Argondizzo, C., and G. Mansfield. 2022. Research on sustainable development literacy and affective learning and teaching actions. *Language Learning in Higher Education* 12, no. 2 (October 26): 365–366. https://doi.org/10.1515/cercles-2022-2066.

Barth, M. 2014. *Implementing sustainability in higher education: Learning in an age of transformation.* London: Routledge. https://doi.org/10.4324/9780203488355.

Berglund, T., N. Gericke, and S.N. Chang Rundgren. 2014. The implementation of education for sustainable development in Sweden: Investigating the sustainability consciousness among upper secondary students. *Research in Science & Technological Education* 32, no. 3 (September 2): 318–339. https://doi.org/10.1080/02635143.2014.944493.

Bianchi, G. 2020. Sustainability competences—A systematic literature review. Luxembourg: Publications Office of the European Union. Available from: https://data.europa.eu/doi/https://doi.org/10.2760/200956 (accessed September 3, 2024).

BMUW/Federal Ministry for the Environment, Nature Conservation, Nuclear Safety and Consumer Protection and German Environment Agency. 2021. *Zukunft? Jugend fragen!—2021: Umwelt, Klima, Wandel—was junge Menschen erwarten und wie sie sich engagieren.* Berlin: Federal Ministry for the Environment, Nature Conservation, Nuclear Safety and Consumer Protection. Available from: https://www.bmuv.de/fileadmin/Daten_BMU/Pools/Broschueren/zukunft_jugend_fragen_2021_bf.pdf (accessed September 3, 2024).

Boeve-de Pauw, J., N. Gericke, D. Olsson, and T. Berglund. 2015. The effectiveness of education for sustainable development. *Sustainability* 7, no. 11 (November 24): 15693–15717. https://doi.org/10.3390/su71115693.

Boyd, F.A. 1991. Business English and the case method: A reassessment. *TESOL Quarterly* 25 (4): 729. https://doi.org/10.2307/3587095.

Brookfield, S. 2012. *Teaching for critical thinking: Tools and techniques to help students question their assumptions*. San Francisco, CA: Jossey-Bass.

Brown, P.C., H.L. Roediger III., and M.A. McDaniel. 2014. *Make it stick: The science of successful learning*. Cambridge, MA: Harvard University Press.

Council of Europe. 2001. *Common European framework of reference for languages: Learning, teaching, assessment*. Strasbourg: Council of Europe Publishing, 2001. Available from: https://rm.coe.int/1680459f97 (accessed September 3, 2024).

Council of Europe. 2020. *Common European framework of reference for languages: Learning, teaching, assessment—companion volume*. Strasbourg:

Council of Europe Publishing. Available from: https://rm.coe.int/common-european-framework-of-reference-for-languages-learning-teaching/16809ea0d4 (accessed September 3, 2024).

de la Fuente, M.J., ed. 2021. *Education for Sustainable Development in foreign language learning*. New York and London: Routledge.

Drugge, A.L. 2021. Inblick i Sámisk historia. In *Inifrån Sápmi. Vittnesmål från stulet land*. Ed. by P. Fjellgren, and M. Nord, 65–77. Stockholm: Verbal Förlag.

European Commission 2020. The European green deal. Available from: https://commission.europa.eu/strategy-and-policy/priorities-2019-2024/european-green-deal_en (accessed September 3, 2024).

Fischer, J., A.M. Abrantes, E. Casey, E. Gigl, and M. Lešnik 2008. *LCas—Language case studies. Teacher training modules for the use of case studies in language teaching at secondary and university level: A handbook.* Strasbourg: European Centre for Modern Languages.

Fischer, J., M.T. Musacchio, and A. Standring. 2009. *EXPLICS—Exploiting internet case studies and simulation templates for language teaching and learning: A handbook.* Göttingen: Cuvillier Verlag Göttingen.

Gabaudan, O. 2022. On a Journey towards education for sustainable development in the foreign language curriculum. *Language Learning in Higher Education* 12, no. 2 (October 26): 525–546. https://doi.org/10.1515/cercles-2022-2060.

Grund, J., and A. Brock. 2020. Education for sustainable development in Germany: Not just desired but also effective for transformative action. *Sustainability* 12, no. 7 (April 2): 2838. https://doi.org/10.3390/su12072838.

Hämäläinen, M. 2022. Ethics, dialogue and English as a lingua franca for ESD in higher education. *Language Learning in Higher Education* 12, no. 2 (October 26): 547–565. https://doi.org/10.1515/cercles-2022-2062.

Kaur, K. 2022. Embed sustainability in the curriculum: Transform the world. *Language Learning in Higher Education* 12, no. 2 (October 26): 605–616. https://doi.org/10.1515/cercles-2022-2061.

Lawrence, R. 2014. Internal colonisation and Indigenous resource sovereignty: Wind power developments on traditional Saami lands. *Environment and Planning D: Society and Space* 32, no. 6 (December): 1036–1053. https://doi.org/10.1068/d9012.

Leal Filho, W., C. Shiel, and A. Paço. 2016. Implementing and operationalising integrative approaches to sustainability in higher education: The role of project-oriented learning. *Journal of Cleaner Production* 133 (October): 126–135. https://doi.org/10.1016/j.jclepro.2016.05.079.

Lozano, R., M. Merrill, K. Sammalisto, K. Ceulemans, and F. Lozano. 2017. Connecting competences and pedagogical approaches for sustainable development in higher education: A literature review and framework proposal. *Sustainability* 9, no. 10 (October 20): 1889. https://doi.org/10.3390/su9101889.

Lyu, E. 2023. Teaching English for specific purposes through case studies: Design, teaching, and evaluation. *Journal of Teaching English for Specific and Academic Purposes* (October 9): 307. https://doi.org/10.22190/JTESAP230225023L.

Maijala, M., N. Gericke, S.-R. Kuusalu, L.M. Heikkola, M. Mutta, K. Mäntylä, and J. Rose. 2024. Conceptualising transformative language teaching for sustainability and why it is needed. *Environmental Education Research*: 1–20. https://doi.org/10.1080/13504622.2023.2167941.

Marcella, V., and Y. Samofalova. 2022. Data-driven and research-based learning approaches to environmental education in university contexts: Two case studies in Italy and Germany. *Language Learning in Higher Education* 12, no. 2 (October 26): 567–586. https://doi.org/10.1515/cercles-2022-2064.

Melin, C.A., ed. 2019. *Foreign language teaching and the environment: Theory, curricula, institutional structures.* New York: The Modern Language Association.

Michelsen, G., H. Grunenberg, C. Mader, and M. Barth. 2016. *Greenpeace Nachhaltigkeitsbarometer 2015—Nachhaltigkeit bewegt die jüngere Generation.* Bad Homburg: VAS-Verlag.

Molina, P. 2022. Bibliographical review on sustainable literacy in language learning and teaching. *Language Learning in Higher Education* 12, no. 2 (October 26): 513–524. https://doi.org/10.1515/cercles-2022-2059.

Nachet, L., C. Beckett, and K.S. MacNeil. 2022. Framing extractive violence as environmental (in) justice: A cross-perspective from Indigenous lands in Canada and Sweden. *The Extractive Industries and Society* 12 (December): 100949. https://doi.org/10.1016/j.exis.2021.100949.

Normann, S. 2021. Green colonialism in the Nordic context: Exploring southern Saami representations of wind energy development. *Journal of Community Psychology* 49, no. 1 (January): 77–94. https://doi.org/10.1002/jcop.22422.

Össbo, Å. 2023. Back to square one. Green sacrifice zones in Sápmi and Swedish policy responses to energy emergencies. *Arctic Review on Law and Politics* 14 (March 17). https://doi.org/10.23865/arctic.v14.5082.

Peltier, J.W., A. Hay, and W. Drago. 2005. The reflective learning continuum: Reflecting on reflection. *Journal of Marketing Education* 27, no. 3 (December): 250–263. https://doi.org/10.1177/0273475305279657.

Sámediggi 2021. *Sápmi—en region som berikar Sverige.* Rennäringspolitisk strategi. Available from: https://www.sametinget.se/156607 (accessed September 3, 2024).

Sámiskt informationscentrum 2024. Karta över Sápmi (map of Sápmi). Available from: https://www.samer.se/karta (accessed September 3, 2024).

Schön, D.A. 1987. *Educating the reflective practitioner: Toward a new design for teaching and learning in the professions.* San Francisco, CA: Jossey-Bass.

Sköld, P. 2011. Development, adjustment and conflict: The Sámi and reindeer husbandry in Sweden in the light of political, social and economic changes. In *Agriculture and forestry in Sweden since 1900: Geographical and historical studies.* Ed. by H. Antonson, and U. Jansson, 475–491. Stockholm: Kungl. Skogs-och Lantbruksakademien.

Sprain, L., and W.M. Timpson. 2012. Pedagogy for sustainability science: Case-based approaches for interdisciplinary instruction. *Environmental Communication* 6, no. 4 (December): 532–550. https://doi.org/10.1080/17524032.2012.714394.

Torres-Goens, L., and L. Farley. 2017. The impact of reflective practice in a university language class. *Reflective Practice* 18, no. 4 (July 4): 448–462. https://doi.org/10.1080/14623943.2017.1307724.

Tuncer, H., and Y. Özkan. 2021. Reflective tools in an English language teaching context: Contributions, challenges and recommendations. *Kahramanmaraş Sütçü İmam Üniversitesi Sosyal Bilimler Dergisi* 18, no. 1 (April 29): 181–204. https://doi.org/10.33437/ksusbd.671482.

Umweltbundesamt/German Environment Agency. 2022. *Umweltbewusstsein in Deutschland 2020: Ergebnisse einer repräsentativen Bevölkerungsumfrage.* Berlin: Federal Ministry for the Environment, Nature Conservation, Nuclear Safety and Consumer Protection (BMUW). Available from: https://www.bmuv.de/fileadmin/Daten_BMU/Pools/Broschueren/umweltbewusstsein_2020_bf.pdf (accessed September 3, 2024).

UNESCO. 2006. Framework for the UN DESD international implementation scheme. Paris: United Nations Educational, Scientific and Cultural Organization. Available from: https://unesdoc.unesco.org/ark:/48223/pf0000148650.locale=en (accessed September 3, 2024).

United Nations. 1973. Report on the United Nations conference on the human environment, Stockholm. New York: United Nations. Available from: https://documents.un.org/doc/undoc/gen/nl7/300/05/pdf/nl730005.pdf (accessed September 3, 2024).

United Nations. 2015. Transforming our world: the 2030 Agenda for sustainable development: The 17 Goals. Available from: https://sdgs.un.org/goals (accessed 3 September 2024).

Vallabh, P. 2018. Youth on the move: Intentions and tensions. In *Issues and trends in Education for Sustainable Development. Education on the move*. Ed. by A. Leicht, J. Heiss, and W.J. Byun, 157–176. Paris: United Nations Educational, Scientific and Cultural Organization. Available from: https://unesdoc.unesco.org/ark:/48223/pf0000261954.locale=en (accessed September 3, 2024).

WCED. 1987. Report of the world commission on environment and development: Our common future. Available from: https://sustainabledevelopment.un.org/content/documents/5987our-commonfuture.pdf (accessed September 3, 2024).

WEF. 2023. *World Economic Forum, innovative learning solutions to navigate complexity: Adapting systems thinking to future classrooms*. White paper. Available from: https://www.weforum.org/publications/innovative-learning-solutions-to-navigate-complexity-adapting-systems-thinking-to-future-classrooms/ (accessed September 3, 2024).

8

Materials for Transformative Language Teaching for Sustainability Model—Students' Perspectives

Minna Maijala, Mareen Patzelt, Päivi Laine, and Soila Merijärvi

Introduction

The discourse that takes place in a language classroom is, to a significant extent, dependent on the teaching materials used in the classroom (Guerrettaz and Johnston 2013; Thoms 2014; Littlejohn 2022). In language classrooms, a certain ecology (van Lier 2004) is built, and learning takes place with the help of affordances, namely, things available in the classroom for learners to do something with. These affordances are often teaching materials (Guerrettaz and Johnston 2013) or the teacher (Thoms 2014) and build relations between learners and the social environment (van Lier 2004). In creating connections between the classroom and the surrounding world, teaching materials can play a crucial role

M. Maijala (✉) · M. Patzelt · P. Laine · S. Merijärvi
School of Languages and Translation Studies, University of Turku, Turku, Finland
e-mail: minna.maijala@utu.fi

M. Patzelt
Åbo Akademi University, Turku, Finland

M. Maijala et al. (eds.), *Transformative Language Teaching for Sustainability*,
https://doi.org/10.1007/978-3-031-85493-4_8

as mediators in promoting awareness and actions to reach Sustainable Development Goals (SDGs). According to the principles of Education for Sustainable Development (ESD), which is the key enabler in reaching the SDGs (see Chapter 1 in this volume), language teaching materials should provide learners with skills in systems-thinking, problem-solving, futures-thinking, and strategic-thinking (Brundiers et al. 2021; Wiek et al. 2011, 2015). ESD in language classrooms should 'give students knowledge, competencies, and tools to live sustainably, and to find solutions to the sustainability challenges we face so that future generations can, too' (de la Fuente 2021, p. 80). The integration of SDGs and ESD into teaching materials is a strategy for equipping students with the knowledge and skills to become informed about SDGs and to become active citizens who contribute to the construction of a more sustainable world. This entails the inclusion of content in language teaching materials that motivates learners to engage in analytical, reflective, and action-oriented learning.

SDG-related content is not directly mentioned in the guiding documents of European language education, which are focused mainly on everyday topics considered helpful for learners communicating in a foreign language (Council of Europe 2001, 2018). The Common European Framework of Reference for Languages (CEFR, Council of Europe 2001, 2020) represents the central framework for the planning of curricula, language textbooks, and other teaching materials. Consequently, the CEFR serves as a guiding document for language educators in the selection of topics for language teaching (Byram 2022). The updated CEFR companion volume (Council of Europe 2020) can be expected to follow the principles of ESD, since it was supplemented with new descriptors for mediation and plurilingual or pluricultural competence. It is becoming increasingly evident that mediation and the ability to function effectively in multilingual environments are crucial competencies for the achievement of SDGs (Maijala et al. 2024). The GreenComp (European Commission 2022) framework, developed by the European Commission for the teaching of all subjects, can also be adapted for use in language teaching and for designing language teaching materials. It includes descriptions of how ESD can be implemented in the following competence areas: *embodying sustainability values, embracing complexity in sustainability*, *envisioning sustainable futures* and *acting for sustainability*. Learner-centred activities, discussions, reflections and problem-solving tasks, and project-based learning are mentioned in the CEFR (Council of Europe 2020), in research literature in the field of ESD (Wiek et al. 2011, 2015; Sinakou et al. 2019; Brundiers et al. 2021), and in the field of language teaching and learning (e.g., Barili and Byram 2021; Maijala et al. 2024).

Language coursebooks vary in the number of topics and exercises that are supposed to promote ESD (see Chapter 6 in this volume). Consequently, there is a need to develop or find supplementary materials for coursebooks to integrate SDG-related content for the purposes of language teaching. The aim of this chapter is to present SDG-related language teaching materials and activities for two language learner groups (Finnish and German as a foreign language) to enable the integration of sustainability topics in language teaching. The set of material was designed for two language courses in two higher education contexts where SDG-related content and activities enhancing ESD were integrated. Additionally, based on a small-scale questionnaire survey, we analysed students' ($N = 17$) reflections on and perceived outcomes from learning with the materials designed for this purpose. The chapter concludes with a discussion of how language teachers can, through adapting the content of language textbooks and development of supplementary materials, contribute to the implementation of SDGs.

Development of Language Teaching Materials

In general, materials for language teaching and learning can be anything that facilitates the learning of a language (Tomlinson 2022). This includes, but is not limited to, books, videos, and mobile applications (Littlejohn 2022). Language teaching materials can inform learners about the target language and culture, guide them in practising the language, provide them with experiences and encouragement in using the target language, and help them make inquiries about the language and culture (Tomlinson 2012). According to Tomlinson (2012, pp. 143–144), materials development 'refers to all the processes made use of by practitioners who produce and/or use materials for language learning, including materials evaluation, their adaptation, design, production, exploitation and research'. Materials development can be characterised as a dynamic process of design, trial, and modification, that is, rewriting the materials in accordance with student receptions of them (Jolly and Bolitho 2011). Within the development process, specific needs of learners should be taken into account (e.g., Amerstorfer and Oxford 2017).

Although teaching materials play an important role in language classrooms (e.g., Guerrettaz and Johnston 2013; Littlejohn 2022), there is a lack of guidance in the research literature and in guidebooks for teachers and student teachers on how to develop materials or adapt existing ones (Masuhara 2022). The research literature on materials development in the field of language teaching focuses mostly on language learning content in general, that is, listening, speaking, reading, writing, vocabulary, grammar, and pronunciation, which can be seen, for instance, in the latest edited volumes on materials development (Bao 2018; Norton and Buchanan 2022). As the integration of SDG-related content has become increasingly important in language education over the past decade (Melin 2019; Goulah and Katunich 2020; de la Fuente 2021; Maijala et al. 2024; see also Chapter 1 in this volume), there is a growing need for guidance on how ESD can be implemented in the context of teaching practice, with language teaching in particular.

In language teaching, teachers often rely on the content of textbooks (Richards 2014). A survey conducted by the British Council in 2008 revealed that 65% of English language teachers worldwide reported utilising textbooks frequently (i.e., *always* or *frequently*) in their teaching practice. Conversely, only 6% of teachers stated that they never used textbooks in their lessons. The central role of textbooks in education has remained largely unchanged over the past few decades, with textbooks continuing to be a central component of language teaching (Rathert and Cabaroğlu 2021; Vitta 2023). For example, the Finnish language education context is characterised by the heavy use of textbooks (Harjanne et al. 2017). Finnish language teachers also use the textbook as a primary source, for instance, for teaching culture (Hahl 2020). In their teaching practice, language teachers should be able to adapt and supplement textbooks to customise their content to the specific learner group and learning context in question (e.g., Grammatosi and Harwood 2013), including for the purpose of ESD implementation.

Implementation of ESD in Language Teaching Materials

Sustainable development (SD) is often divided into three dimensions: social, environmental, and economic (e.g., Giddings et al. 2002). In the research literature, the cultural dimension has been identified as a dimension of its own, alongside the social dimension (e.g., Burford et al. 2013). The social dimension, including the practice of democratic skills through interaction with others, has long been included in language teaching (Barili and Byram 2021). The focus of language teaching is on interaction, that is, on developing learners' skills to use the target language in a communicative way (Bachman and Palmer 2010). Thus, due to the special features of the subject, the main focus of language teaching for SD lies usually on its cultural and social dimensions (Maijala et al. 2023), emphasising cultural specificities and interactive activities (Hiver et al. 2021). In language textbooks, topics related to the cultural and social dimensions of the target country are often addressed (Mohammadnia and Moghadam 2019). Interactive working methods are central to language teaching, which can support the practice of democratic skills (e.g., Wagner et al. 2018) and the implementation of ESD (e.g., Sinakou et al. 2019). In an article by Müller (2022), communicative teaching activities are proposed as a means of introducing and developing ESD-relevant skills and of integrating ESD into the teaching of German as a foreign language. These include, for instance, interactive role-playing and vocabulary games, panel discussions, and a sustainability diary. Some of these were adapted for the present study.

When implementing ESD, finding and developing suitable materials is a challenge for teachers (Waltner et al. 2020; Letouzey-Pasquier et al. 2023). For instance, a study by Rieß and others (2008) reported that teachers who integrated ESD-related content in their lessons had better knowledge about ESD teaching materials compared to teachers who had not taught ESD-relevant content. A study by Mercer and others (2023) revealed that 74.7% of language teachers ($N = 91$) reported not using coursebooks when dealing with environmental issues in their lessons. Most teachers (75/91) stated that they used online materials or produced materials themselves (57/91). In general, language teachers tend to prefer ready-made textbooks over developing materials themselves (Carabantes and Paran 2022). As far as we know, information on language teachers implementing ESD is mainly available from studies involving teachers of many different subjects (e.g., Uitto and Saloranta 2017; Ferguson et al. 2021; Sund et al. 2020). These studies showed that language teachers integrated SDG-related content in their lessons, primarily through various media (Sund et al. 2020). Language teachers' motivation and attitudes seem to influence their implementation of ESD (Kwee 2021). From the student perspective, the integration of SDG-related content and language teaching seems to be beneficial (e.g., de la Fuente 2021). Studies (e.g., Alcantud-Díaz and LLoret-Catalá 2023; Sommier et al. 2022) have demonstrated that university students' awareness of sustainability topics is enhanced when they attend courses that include such topics.

Valnes Quammen and Afitska (forthcoming) describe the integration of sustainability-related content into language courses in non-language departments at the advanced level in a US higher education context. They also show how existing materials can be adapted for ESD implementation in language courses. Adaptations consist of more elaborate thematic vocabulary banks, redesigning tasks to include students' production of targeted structural features, and then presenting their work. They also present how key competencies in sustainability (Brundiers et al. 2021; Wiek et al. 2011, 2015) can be added to existing materials that do not offer explicit links between linguistic and sustainability-related content and sustainability competencies. In their adaptations, interpersonal competency methods were supported through guided small-group work and collaborative work compiling of multimodal case studies, which were presented in small groups. Another ESD competence, namely values-thinking, was added to existing materials by exploring historical and cultural implications for sustainability problems and by comparing student communities to Francophone communities. The process of adapting and supplementing textbooks with content related to SDGs can serve as an empowering aspect for language teachers (see Bouckaert 2019).

Theories in the field of language teaching and learning have been adapted especially in the multidisciplinary model of *Transformative Language Teaching for Sustainability* (TLS, Maijala et al. 2024; see Chapter 1 in this volume). The TLS model aims to help language teachers identify topics and teaching methods supporting the integration of SDGs and ESD into language teaching practice. The model takes into account both the content of teaching and teaching methods. At the content level, the TLS model encourages including, besides 'traditional' language learning content related to the target culture, topics such as the environment and climate change, and other related topics that encourage learners to take responsibility for the environment. According to the model, methods of language teaching that are already widely used can be combined with teaching methods that have been found to be beneficial in the research literature on sustainability education (Sinakou et al. 2019). These are learner-centred teaching methods that emphasise action, increase learners' change-agency skills, and develop competences such as conflict management, decision-making, and problem-solving. The TLS model brings together learner-centred activities, such as discussions and problem-solving tasks, which are already highlighted also in the CEFR (Council of Europe 2001, 2020).

This chapter describes the development of language teaching materials based on a theoretical model (Maijala et al. 2024; see Chapter 1 in this volume) that includes SDG-related content and follows the principles of ESD with a focus on student perspectives. It also analyses how these materials were used in two language learning settings in higher educational contexts. The objective of the study reported here was to examine the practical application of the theory and the subsequent impact of the use of the materials. This entailed translating theoretical concepts into practical outcomes (McGrath 2002; Timmis 2022). To address the overall aim of developing teaching materials for the achievement of SDGs by implementing ESD, we address our research question as follows: How did students experience working with the materials designed to promote ESD?

Context of the Study

The aims of the research project Ethical and Sustainable Language Teaching (EKKO 2021–2025) were to examine the role of SD and how it can be promoted in language teaching and to develop engaging language learning activities promoting ESD. We designed language teaching materials based on the TLS model (Maijala et al. 2024) and ESD research literature (Brundiers et al. 2021; Sinakou et al. 2019; Wiek et al. 2011, 2015; see Fig. 8.1). All materials and activities were linked to SDGs. By incorporating the TLS model and ESD methods into the materials design, our aim was to encourage students' active participation and critical thinking by developing activities that allow them to explore real-world issues related to the SDGs (Fig. 8.1). Student engagement (discussions, group work, problem-solving, and project-based learning) was regarded as an important goal in the materials development process. The aim was to highlight the global significance of SDGs while at the same time emphasising their relevance to local communities. This happened by connecting the materials' goals to both global and local communities and environments.

Figure 8.1 shows the framework for the developed materials building on the TLS model (Maijala et al. 2024) and ESD research literature (Brundiers et al. 2021; Sinakou et al. 2019; Wiek et al. 2011, 2015). Similarly to Wiek and others (2011, p. 204), we define competence as 'a functionally linked complex of knowledge, skills, and attitudes that enable successful task performance and problem solving'. As defined by Wiek and others, key competences for sustainability in higher education serve the goal of enabling students to 'plan, conduct, and engage in sustainability research and problem-solving based on the interplay of systems-thinking, anticipatory, normative, strategic, and interpersonal competencies' (Wiek et al. 2011, p. 207). Interpersonal competencies, as described by Brundiers and others (2021), are essential skills for effective

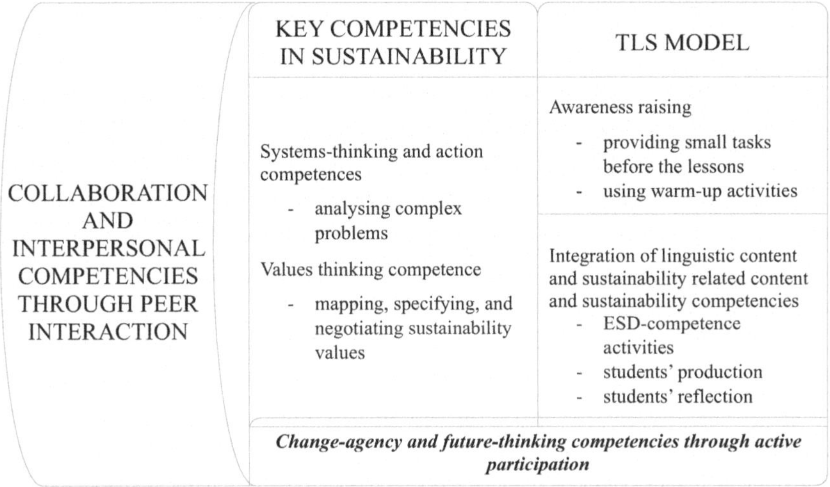

Fig. 8.1 Framework for the developed materials building on the Transformative Language Teaching for Sustainability (TLS) model (Maijala et al. 2024) and Education for Sustainable Development (ESD) research literature

collaboration and communication in promoting ESD. These competencies include the ability to work with others and manage conflicts constructively. The importance of empathy, active listening, and the capacity to build and maintain relationships is emphasised, as these are essential when addressing complex SDG-related issues. For our analysis, we also added the aspect of awareness as being part of the competencies we looked to identify in participant reflections (Maijala et al. 2024). Awareness can be seen to affect and existentially belong to knowledge, skills, and attitudes one holds or obtains (Fantini 2020). The set of four areas (awareness, attitudes, skills, and knowledge) was originally suggested by Fantini (2020) to describe aspects of intercultural communicative competence one obtains, besides academic learning, through experiential learning as well as discussion and reflection. As key components of competencies, however, we saw these four areas also as fit to our purpose of investigating the competencies participants gained in our courses. Through identifying more abstract competencies in participant reflections, we aimed to draw conclusions on the more concrete learning outcomes for the participants (see Wiek et al. 2011).

This chapter describes two language courses, Finnish as a foreign language (FFL) and German as a foreign language (GFL) in higher education that were designated to incorporate SDG-related content and ESD teaching methods into language teaching (see Table 8.1). In the FFL group, students of Finnish as a foreign language from various German-speaking universities took part in a three-hour onsite course that was held for the main part in German. Students' knowledge of Finnish varied from the beginner level to a very high level (ranging from A2 to C2 on the CEFR scale). The aim of the course was to learn and revise vocabulary needed to discuss nature and reflect on its importance for humans in the target language. The GFL group, students of German as a foreign language from three European universities, attended a five-day language course in higher education as part of an international master's programme. The objective of the course was to demonstrate how SDGs can be integrated into the teaching of German as a foreign language. The course's language learning objectives were twofold: firstly, to enhance oral language proficiency, and secondly, to foster intercultural communication and comprehension. The language of instruction was German because students' language levels were high (ranging from B1 to C2 on the CEFR scale).

SDG-related materials for both language courses were developed based on the framework depicted in Fig. 8.1. This was done also keeping in mind students' skills in the language. The activities are described in detail in Table 8.1. In the FFL group, students worked on the following SDG-related activities: (1) preliminary work, (2) short group tasks on four topics, and (3) the presentation of group tasks and collective final discussion. In the GFL group, the materials and activities included (1) preliminary work, (2) keeping a sustainability diary, (3) attending group work on examples of local and global SDGs, and (4) discussing in a panel the advantages and disadvantages of the possibilities to achieve SDGs. Students prepared for the courses differently. In the FFL group, students worked independently on different activities before the course. In the GFL group, students participated in online meetings, where various group activities were carried out.

Subsequent to both courses, students completed an online questionnaire (see Appendix 2). Besides background information, students

Table 8.1 Materials and activities related to Sustainable Development Goals in Finnish and German as a foreign language groups

FFL group (Finnish as a foreign language)	GFL group (German as a foreign language)
Preparations before the course Students went on walks in nature finding words in Finnish related to forest, waters, and birds and reflected on what they like to do in nature.	*Preparations before the course* Students were prepared for the onsite course during online meetings. During these, students worked together on different SDG-related activities, such as presenting student life and their home environment, imagining their future home environment, and writing an elf tale (*Elfchen*) on the subject of 'forest' or 'nature'.
Course activities Instructions for station learning (4 stations) In groups of 4, students go through four stations (20 min/station). → In the end, each group presents (in 3 minutes) the station they visited last, and other groups comment. *Baltic Sea* • (Warming up) Pictures symbolising issues the Baltic Sea currently faces: (plastic) waste, ship traffic for trade and leisure, decline in biodiversity, eutrophication). → Students collect words in Finnish and/or German for what they see and what they associate with the pictures. → Small-group discussion • (Activity) Students watch a video about the dangers the inhabitants of the Baltic Sea region face. → Students collect their own ideas and use information from the video to answer questions on the future of the Baltic Sea and potential dangers for it as well as what humans can do to alleviate problems/protect the Baltic Sea in the future. → Answers are written into a table in Finnish and/or German • (Reflection) Students discuss why it is important for us as humans to protect the Baltic Sea (and waters in general) and note their ideas on a shared poster for this station.	*Course activities* *Sustainability diary (adapted from* Müller 2022) • (Warming up) Students keep daily diary entries during the course (notebook, audio recordings) in which they document their activities, observations, and thoughts on the topic of sustainability during the 5-day course. • (Activity) Students are asked to describe the sustainable activities they implement during the course, such as using public transport, avoiding single-use plastic, or taking part in environmentally friendly activities. • (Reflection) At the end of the course, students are asked to reflect in groups on their personal challenges, successes, and insights with regard to sustainable behaviour.

(continued)

Table 8.1 (continued)

FFL group (Finnish as a foreign language)	GFL group (German as a foreign language)
Birds • (Warming up) Students look at pictures of birds (with the bird's name in Finnish) and collect adjectives to describe the birds. → They make a list within their small group. • (Activity) Cards are laid out on the table. One student at a time describes one of the birds without naming it, and other students guess which bird is described. • (Reflection) Students discuss possible threats birds in Finland face (e.g., climate crisis) and other dangers for birds in Finland. → Note ideas in Finnish and/or German on shared poster for this station. *Moving around in nature* • (Warming up) Students plan their dream outing into (Finnish) nature, think of activities they would like to do and plan the itinerary. → Notes are in Finnish and/or German. • (Activity) Students discuss challenges tourism can bring for nature. → They write rules in Finnish on how one should behave when in nature (they are provided with notes on the use of imperative in Finnish and constructions with *One may (not)...*). • (Reflection) Students collect ideas and discuss what responsible tourism can mean. → Notes in Finnish and/or German placed on a shared poster for this station. *Forest* • (Warming up) Students watch a video of a walk through in a forest, narrated in Finnish. • (Activity) Students collect and note words they remember from the video and all other words they associate with forests that they know in Finnish and/or German. → Students draw a mind map and sort/group the words around the question: Why is the forest important for me and for humans in general? • (Reflection) Students discuss what possible dangers forests face nowadays and what humans can do to minimise these risks.	*Group work: global vs. local goals* • (Warming up) Students find an example of a local and a global SDG, e.g., their hometown vs. the world (see Appendix 1). • (Activity) Students discuss in groups about the local and global goals and prepare a poster. The posters are assembled into an exhibition. • (Reflection) During the exhibition, participants are invited to identify the personal goals that are of greatest importance to them. *Panel discussion (adapted from* Müller 2022) • (Activity) Students are divided into three groups. Group 1 collects arguments as to why the SDGs might not be realisable or only with difficulty. Group 1 prepares a short speech against the SDGs and makes it clear why the goals contradict their own interests and should therefore not be observed/implemented by the population. Group 2 collects arguments as to why the SDGs are achievable and prepares a short speech in favour of the SDGs and explains why it is important for the population to achieve the goals. Group 3 collects information about the feasibility of the SDGs. • (Reflection) After the discussion between groups 1 and 2, group 3 takes the role of experts and assesses the persuasiveness of both groups.

answered questions regarding their expectations of the course, their understanding of sustainability, their preferred ways of learning a language, and what they learned during the course. Altogether, 17 students from various universities in Austria, Finland, Germany, and Spain participated in the survey. The students who participated had the following first languages: German (10), Finnish (3), Italian (1), Kurdish (1), Kirghiz (1), or Russian (1). One student did not mention their first language and one student was bilingual mentioning two first languages. Participants studied various major and minor subjects at university, such as Finnish studies / Finnish as a foreign or second language / Finno-Ugric studies (10), German as a foreign or second language / German studies (6), English studies (4), Scandinavian studies (2), history (2), literature (2), linguistics (1), Arabic studies (1), business administration (1), Dutch philology (1), Indogermanic studies (1), Italian (1), language, literature, and culture studies (1), Spanish (1), translation (1), and political science. The questions were asked and answered in German; the authors translated the answers into English. The students were informed that they were participating in a research study, and the purpose of the study was described to them (Wagner 2015). A privacy notice was included on the cover of the questionnaire informing the participants of the use of research data. The participants were asked for permission to use the material for research.

Research Methods

For the analysis, we adopted a qualitative approach for the responses to the questionnaire (see Appendix 2), and, in the first step of the analysis (see Table 8.2), identified topics within participants' reflections we saw as belonging to one of the four areas of awareness, attitudes, skills, and knowledge.

In the second step of the analysis, we therefore aimed to identify key competences for sustainability in the participants' reflections through assigning what we identified as awareness, attitudes, skills, and knowledge to the key competences in sustainability as defined by Wiek and others (2011) (see Table 8.2 for short descriptions). This was done by

Table 8.2 Two-step analysis of questionnaire responses

The first step of the analysis				
Identifying competence components (awareness, attitudes, skills, knowledge)				
The second step of the analysis				
Assigning competence components to key competences for sustainability ⇩				
Key competences for sustainability (Wiek et al. 2011, pp. 207–211)				
Systems-thinking competence	Anticipatory competence	Normative competence	Strategic competence	Interpersonal competence
Comprehending and analysing complex systems in sustainability and the dynamics within	Constructing, comprehending, and evaluating images of the future in relation to sustainability	Mapping, applying and negotiating values, principles, and goals in sustainability	Understanding strategic concepts and knowing how to take action for transition towards sustainability	Understanding, comparing, and critically evaluating different perspectives
Building transition strategies based on acquired systemic knowledge	Anticipating and preventing harmful consequences to sustain intergenerational equity	Assessing the state of social-ecological systems and creating sustainability visions for them	Addressing sustainability's key issues through practical solutions for complex problems	Collaborating towards solving sustainability problems while understanding and facilitating diversity

close reading of the participants' answers to questions about what they had learned in the course and what they would take home from it.

Results

The answers gave a general insight into how the students defined sustainability for themselves. When students were asked about their understanding of sustainability, 14 out of 17 explicitly mentioned ecological sustainability. We would like to note that the course in the German-speaking university was focused on the ecological dimension specifically, and the questionnaire was completed at the end of the course. This may have had an influence on the understanding of the concept. Some elaborated on this, while others simply stated this dimension to be part of their understanding of SD. These students included in their answers the following general topics: nature as a resource that should be protected and used responsibly; consequences of human activity for nature and wildlife; the threat of climate change and awareness of climate protection; and ways to help preserve nature and protect the climate for future generations. Two students explained their understanding of sustainability without specifically elaborating on any of the dimensions. One student defined SD as development which takes into consideration its possible consequences. Another student described how sustainable development to them meant having an understanding of sustainability (and its purpose) regarding different aspects. One answer defined sustainability as a way to preserve the environment, economy, and society so that they keep functioning and provide conditions worth living in also in the future.

Regarding the participants' reflections on their experience working with the materials designed to promote ESD and what they learned in the course, we identified awareness, attitudes, skills, and knowledge within all of the key competences described above (Table 8.3).

Table 8.3 Competence components (attitudes, skills, knowledge, and awareness) in participant reflections in the key competences for sustainability

Key Competences	Competence components			
	Attitude	Skill	Knowledge	Awareness
Systems-thinking competence	Connections between elements within the ecosystem that should be protected Importance of integrating sustainability into all levels of education	Integrating sustainability and ecological awareness into language education Generally taking into consideration ecological, economic, and social aspects of sustainability	Challenges for the ecosystem caused by humans and/or climate change Origin of products we consume	Relevance of current urgent problems
Anticipatory competence			Long-term solutions for sustainability on a societal/political level Protecting the environment	(In)visibility of consequences for one's own actions
Normative competence	Shared value of protecting nature / treating nature responsibly and mindfully Language education as one possible forum to address ecological topics	Reflecting and developing one's own behaviour regarding sustainability	Defining sustainable education SDGs	Importance of nature as a multifaceted benefactor of well-being and prosperity Raising awareness for ecological challenges

Key Competences	Competence components			
	Attitude	Skill	Knowledge	Awareness
Strategic competence	Willingness to act for climate protection as an individual	Integrating sustainability and climate protection into everyday life through tangible individual actions	Adopting sustainable principles for interacting with the world as an ecosystem Short-term solutions on an individual level	Individual action for protecting nature Tailored solutions for diverse problems
Interpersonal competence	Excitement about similar opinions and variety of perspectives among course participants	Reflecting one's own culture, values and actions Reducing prejudices Intercultural communication Teamwork Discussing sustainable development	Vocabulary for sustainability in Finnish Knowledge through others' expertise and about others' backgrounds around sustainability	Individual differences in perceptions of one's relation to nature

Topics identified in participants' answers were relatively evenly distributed among four of the five key competences for sustainability defined by Wiek and others (2011). In general, topics extracted from the participants' reflections contain tangible examples of the way they conceptualised sustainability in or following the courses.

Participants' answers showed different attitudes, skills and knowledge that can be attributed to *systems-thinking competence*. Their answers included remarks on the connections between various dimensions and scales of sustainability. They also mentioned how addressing sustainability topics in, for example, language learning and understanding the relevance of current issues can contribute to systemic knowledge and to building transition strategies.

Topics attributed to *anticipatory competence* in participants' reflections addressed long-term solutions for sustainability on a societal or political level, environmental protection in general, and thoughts about the consequences of one's own actions. Participants mentioned attitudes and skills within anticipatory competence comparatively rarely. One reason for this might be that constructing and evaluating images of a sustainable future was at the core of both courses we delivered. Students thus mentioned more concrete examples that could (more or less) all be seen to also fit the description of anticipatory competence, as they generally circled around constructing and comprehending an image of a sustainable future and preventing harmful consequences. General evidence for this competence was also apparent in their definitions of sustainability, which some explicitly linked to liveable conditions for future generations.

Within *normative competence*, we identified, for example, topics linked to acknowledging and validating one's own and others' values of treating nature respectfully, knowledge about the definition of sustainability and SDGs, and awareness of nature as a multifaceted benefactor of individual and societal well-being and prosperity. Participants showed in their answers that they can assess and reflect on the state of the social-ecological system we live in and, through reflecting on their own and shared values, construct a sustainability vision.

Strategic competence in participants' reflections showed, for example, in descriptions of concrete actions for transition towards sustainability. These included solutions on an individual level, such as integrating sustainability into one's everyday life and in occasional activities like travelling, rules for being in nature and ways to protect it, and working for climate protection individually and as a society. Participants also mentioned that problems may be complex and that possible solutions should be tailored to each individual problem.

Participants also compared different perspectives that came up during the courses, which we linked to their *interpersonal competence*. In their reflections, they appreciated the diversity in perspectives as well as similarities they discovered between themselves and others. They also reported that they critically evaluated their own culture and values and reduced prejudices. Teamwork, intercultural communication, and discussion skills were mentioned as contributing to solving sustainability. Language skills were also closely linked to this.

Discussion and Conclusion

The results of the study show that through tailoring suitable material for courses addressing SD within language education, participants acquired a variety of competences. We can conclude that students acquired new vocabulary in general and specifically within the topic of nature that they activated and practised already existing language skills. Participants learned to express adequately and to share in the target language what they had partly already known about sustainability from their general education. Students also acquired new factual knowledge about sustainability and nature and developed their thinking skills. To sum up, combining SDG-related topics and ESD methods supported the language learning process.

The most beneficial activities were those that facilitated interaction and reflection on the role of SD in the students' own environment, as well as comparison with their experiences in the target country. The task around the topic of the forest, for instance, prompted lively discussions among the students of Finnish, touching upon various aspects of how the forest is important to us as a society and the students personally, and connecting these aspects to one another. Students stressed the importance of trees for producing oxygen and the forest as an important factor for biodiversity. They also described the value of forests as cultural heritage and for (responsible) commercial use. On an individual level, benefits for mental health came up in the discussion.

To answer the research question, 'How did students experience working with the materials designed to promote ESD?', we conducted a small-scale questionnaire study among language students in two higher education settings. We clearly identified five key competences for sustainability (Brundiers et al. 2021; Wiek et al. 2011, 2015). Students were able to make connections between sustainability dimensions, which refers to systems-thinking. Anticipatory competence, closely linked to systems-thinking (Sinakou et al. 2019), could be detected in students' understanding of long-term solutions for sustainability and in their reflection on the consequences of one's own actions. In students' answers, we identified topics related to normative competence, such as validating one's own and others' values of treating nature respectfully and increased awareness of nature. We discerned reflections showing strategic competence, for instance, in descriptions of concrete action for sustainability. Reflections related to interpersonal competence included appreciation of diversity. Language skills were linked to interpersonal competence.

Even though the main focus in language education is on language and the development of language and communication skills, our case study has shown that suitable materials can also help integrate sustainability topics into language teaching. Similar materials could be adapted globally to various languages in multiple teaching contexts. Appropriate tasks for the level and age of learners can help promote holistic understanding of sustainability topics and prompt learners to acquire attitudes, skills, knowledge, and awareness necessary to discuss, reflect, and implement SDGs on local and global levels as well as on personal and societal levels. Language education can especially contribute to interpersonal competencies for sustainability, as it can promote dialogue and mutual understanding across languages and cultural backgrounds (Hiver et al. 2021).

Despite the small number of respondents in this study, which prevents broader generalisations, our findings suggest that students benefited from the incorporation of SDG-related content in the language classroom. They found that their language skills improved, and, at the same time, their awareness of SDGs increased. Further research is required to gain insight into how the development and adaptation of teaching materials in alignment with SDGs can empower teachers (Bouckaert 2019), as well as the specific professional knowledge and skills that language teachers require to achieve this.

Appendix 1

Activity in German as a Foreign Language Group: Global vs. Local Goals

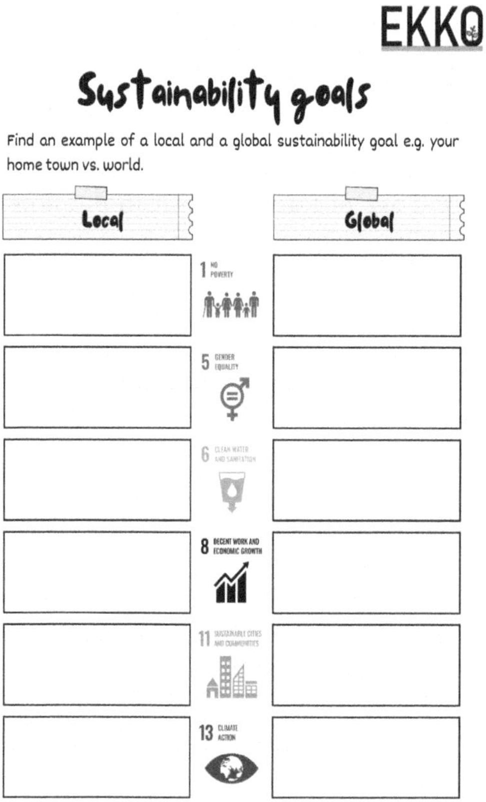

Appendix 2

Questionnaire

[The original questionnaire was administered in German. The questionnaire was translated from German to English by the authors]

1–4. Basic information (year of birth; L1; major and minor subject(s) at the university).
5. What do you understand by sustainable development?
6. What were your expectations of this course?
7. Please briefly describe what you have learnt in the course.
8. What do you take with you from the course into your everyday life?
9. Which of the working methods listed below do you think is best for learning languages, either in the classroom or outside the classroom? You can choose a maximum of three alternatives.

Project-based work
Group work
Working independently
Working in pairs
Interactions with artificial intelligence
Lecture
Others, which?

10. Which of the exercises or tasks listed below do you think you learn languages best with, either in the classroom or outside the classroom? You can choose a maximum of three alternatives.

Simulations and role plays
Oral activities/discussions in the target language
Self-evaluation and reflection
Creative exercises
Practical exercises
Listening exercises
Reading exercises

(continued)

(continued)

Vocabulary exercises
Writing exercises
Practising the target language with native speakers
Language learning apps
Competitions and games
Preparing and giving presentations
Grammar exercises
Others, which?

11. You are allowed to use my data and materials from the course in an anonymised form for research purposes.

Yes
No

References

Alcantud-Díaz, M., and C. LLoret-Catalá. 2023. Bridging the gap between teacher training and society. Sustainable Development Goals (SDGs) in English as a foreign language (EFL). *Globalisation, Societies and Education*, 1–12 (November 20). https://doi.org/10.1080/14767724.2023.2283510

Amerstorfer, C.M., and R.L. Oxford, ed. 2017. *Language learning strategies and individual learner characteristics: Situating strategy use in diverse contexts.* London: Bloomsbury Academic.

Bachman, L.F., and A. Palmer. 2010. *Language assessment in practice: Developing language assessments and justifying their use in the real world*. Oxford: Oxford University Press.

Bao, D., ed. 2018. *Creativity and innovations in ELT materials development: Looking beyond the current design.* Bristol: Multilingual matters.

Barili, A., and M. Byram. 2021. Teaching intercultural citizenship through intercultural service learning in world language education. *Foreign Language Annals* 54, no. 3 (October): 776–799. https://doi.org/10.1111/flan.12526.

British Council. 2008. *Teaching English: Course books*. London: British Council.

Bouckaert, M. 2019. Current perspectives on teachers as materials developers: Why, what, and how? *RELC Journal* 50, no. 3 (December): 439–456. https://doi.org/10.1177/0033688218810549

Brundiers, K., M. Barth, G. Cebrián, M. Cohen, L. Diaz, S. Doucette-Remington, W. Dripps, G. Habron, N. Harré, M. Jarchow, K. Losch, J. Michel, Y. Mochizuki, M. Rieckmann, R. Parnell, P. Walker, and M. Zint. 2021. Key competencies in sustainability in higher education—Toward an agreed-upon reference framework. *Sustainability Science* 16, no. 1 (January): 13–29. https://doi.org/10.1007/s11625-020-00838-2

Burford, G., E. Hoover, I. Velasco, S. Janoušková, A. Jimenez, G. Piggot, D. Podger, and M. Harder. 2013. Bringing the "missing pillar" into Sustainable Development Goals: Towards intersubjective values-based indicators. *Sustainability* 5, no. 7 (July 12): 3035–3059. https://doi.org/10.3390/su5073035.

Byram, M. 2022. Politics, origins and futures of the CEFR. *The Language Learning Journal* 50, no. 5 (September 3): 586–599. https://doi.org/10.1080/09571736.2020.1845392

Carabantes, L., and A. Paran. 2022. 'I preferred to take another activity from the textbook': An activity–theoretical study of learning to design language teaching materials. *The Modern Language Journal* 106, no. 4 (December): 659–674. https://doi.org/10.1111/modl.12803

Council of Europe. 2001. *Common European framework of reference for languages: Learning, teaching, assessment*. Strasbourg: Council of Europe. https://rm.coe.int/1680459f97 (accessed June 6, 2024).

Council of Europe. 2020. *Common European framework of reference for languages: Learning, teaching, assessment—Companion volume*. Council of Europe Publishing, Strasbourg. Available from: www.coe.int/lang-cefr. (accessed June 6, 2024).

de la Fuente, M.J. 2021. Problem-based pedagogy for the advanced college language curriculum: Becoming a multicompetent language user through sustainability education. In *Education for Sustainable Development in foreign language learning: Content-based instruction in college-level curricula*, 65–84. 1st ed. Routledge Research in Language Education. New York: Routledge.

EKKO. 2021–2025. *Ethical and sustainable language teaching*. https://sites.utu.fi/ekko/en/ (accessed July 2, 2024).

Fantini, A.E. 2020. Reconceptualizing intercultural communicative competence: A multinational perspective. *Research in Comparative and International Education* 15, no. 1 (March): 52–61 https://doi.org/10.1177/1745499920901948

Ferguson, T., C. Roofe, and L.D. Cook. 2021. Teachers' perspectives on sustainable development: The implications for Education for Sustainable Development. *Environmental Education Research* 27, no. 9 (May 6): 1343–1359. https://doi.org/10.1080/13504622.2021.1921113

Giddings, B., B. Hopwood, and G. O'Brien. 2002. Environment, economy and society: Fitting them together into sustainable development. *Sustainable Development* 10, no. 4 (November): 187–196. https://doi.org/10.1002/sd.199.

Goulah, J., and J. Katunich, eds. 2020. *TESOL and sustainability: English language teaching in the Anthropocene era*, 1st ed. London: Bloomsbury Publishing.

Grammatosi, F., and N. Harwood. 2013. An experienced teacher's use of the textbook on an academic English course: A case study. In *English language teaching textbooks: Content, consumption, production*, ed. N. Harwood, 178–204. Basingstoke: Palgrave Macmillan UK.

European Commission. Joint Research Centre. 2022. *GreenComp: The European sustainability competence framework.* https://data.europa.eu/doi/10.2760/13286 (accessed July 2, 2024)

Guerrettaz, A.M., and B. Johnston. 2013. Materials in the classroom ecology. *The Modern Language Journal* 97, no. 3 (September): 779–796. https://doi.org/10.1111/j.1540-4781.2013.12027.x

Hahl, K. 2020. Kulttuurin muuttunut olemus opetussuunnitelmissa ja vieraiden kielten oppikirjoissa [The transformation of culture in curricula and foreign language textbooks]. In *Kielididaktiikan katse tulevaisuuteen: Haasteita, mahdollisuuksia ja uusia avauksia kielten opetukseen*, eds R. Hilden and K. Hahl, 173–201. Helsinki: University of Helsinki.

Harjanne, P., C.D. Larenas, and S. Tella. 2017. Foreign-language teaching and studying in Chilean and Finnish classrooms as seen by teachers. *Journal of Language and Cultural Education* 5, no. 3 (September 1): 1–21. https://doi.org/10.1515/jolace-2017-0025

Hiver, P., A.H. Al-Hoorie, J.P. Vitta, and J. Wu. 2021. Engagement in language learning: A systematic review of 20 years of research methods and definitions. *Language Teaching Research* (March 24): 136216882110012. https://doi.org/10.1177/13621688211001289

Jolly, D., and R. Bolitho. 2011. A framework for materials writing. In *Materials development in language teaching*, 2nd ed., ed. B. Tomlinson, 107–134. Cambridge: Cambridge University Press.

Kwee, C.T.T. 2021. I want to teach sustainable development in my English classroom: A case study of incorporating Sustainable Development Goals in

English teaching. *Sustainability* 13, no. 8 (April 9): 4195. https://doi.org/10.3390/su13084195

Letouzey-Pasquier, J., B. Gremaud, S. Blondin, and P. Roy. 2023. Development of teachers' practices in the field of Education for Sustainable Development (ESD): A discursive community of interdisciplinary practices focusing on the theme of chocolate. *Environmental Education Research* 29, no. 8 (August 3): 1155–1169. https://doi.org/10.1080/13504622.2022.2128056

Littlejohn, A. 2022. The analysis and evaluation of language teaching materials. In *The Routledge handbook of materials development for language teaching*, ed. J. Norton and H. Buchanan, 263–276. London: Routledge.

Maijala, M., L.M. Heikkola, S.-R. Kuusalu, P. Laine, M. Mutta, and K. Mäntylä. 2023. Pre-service language teachers' perceptions of sustainability and its implementation in language teaching. *Language Teaching Research* (May 13). https://doi.org/10.1177/13621688231170682

Maijala, M., N. Gericke, S.-R. Kuusalu, L.M. Heikkola, M. Mutta, K. Mäntylä, and J. Rose. 2024. Conceptualising transformative language teaching for sustainability and why it is needed. *Environmental Education Research* 1–20. https://doi.org/10.1080/13504622.2023.2167941.

Masuhara, H. 2022. Approaches to materials adaptation. In *The Routledge handbook of materials development for language teaching*, ed. J. Norton and H. Buchanan, 277–283. London: Routledge.

McGrath, I. 2002. *Materials evaluation and design for language teaching*. Edinburgh: Edinburgh University Press.

Melin, C., ed. 2019. *Foreign language teaching and the environment: Theory, curricula, institutional structures*. Teaching languages, literature, and culture. 6. New York: The Modern Language Association of America.

Mercer, S., N. Correia Ibrahim, K. Bilsborough, C. Jones, and C. Potzinger. 2023. Teacher perspectives on addressing environmental issues in ELT. *ELT Journal* 77, no. 4 (December 12): 393–406. https://doi.org/10.1093/elt/ccac039

Mohammadnia, Z., and F.D. Moghadam. 2019. Textbooks as resources for Education for Sustainable Development: A content analysis. *Journal of Teacher Education for Sustainability* 21, no. 1 (June 1): 103–114. https://doi.org/10.2478/jtes-2019-0008

Müller, D. 2022. Nachhaltigkeit im Fremdsprachenlernen: Ansätze für den universitären DaF-Unterricht unter der Zielsetzung einer Bildung für nachhaltige Entwicklung (BNE). *Informationen Deutsch Als Fremdsprache* 49, no. 1: 93–116.

Norton, J., and H. Buchanan, eds. 2022. *The Routledge handbook of materials development for language teaching*, 1st ed. London: Routledge.

Rathert, S., and N. Cabaroğlu. 2021. Teachers as slaves or masters to their coursebooks: An in-depth study on two English language teachers' coursebook utilization. *Language Teaching Research* 28, no. 5 (September): 1816–1841. https://doi.org/10.1177/13621688211036239

Rieß, W., C. Mischo, A. Reinholz, K. Richter, C. Dobler, and H. Seybold. 2008. *Evaluationsbericht Bildung für Nachhaltige Entwicklung an weiterführenden Schulen in Baden-Württemberg*. Germany: Aktionsplan Baden-Württemberg.

Richards, J.C. 2014. The ELT textbook. In *International perspectives on materials in ELT*, ed. K. Graves, 19–36. Basingstoke: Palgrave Macmillan.

Sinakou, E., V. Donche, J. Boeve-de Pauw, and P. Van Petegem. 2019. Designing powerful learning environments in Education for Sustainable Development: A conceptual framework. *Sustainability* 11, no. 21 (October 28): 5994. https://doi.org/10.3390/su11215994

Sommier, M., Y. Wang, and A. Vasques. 2022. Transformative, interdisciplinary and intercultural learning for developing HEI students' sustainability-oriented competences: A case study. *Environment, Development and Sustainability* (March 1). https://doi.org/10.1007/s10668-022-02208-7

Sund, P., N. Gericke, and G. Bladh. 2020. Educational content in cross-curricular ESE teaching and a model to discern teacher's teaching traditions. *Journal of Education for Sustainable Development* 14, no. 1 (March): 78–97. https://doi.org/10.1177/0973408220930706

Thoms, J.J. 2014. An ecological view of whole-class discussions in a second language literature classroom: Teacher reformulations as affordances for learning. *The Modern Language Journal* 98, no. 3 (September): 724–741. https://doi.org/10.1111/modl.12119

Timmis, I. 2022. Theory and practice in materials development. In *The Routledge handbook of materials development for language teaching*, ed. J. Norton, and H. Buchanan, 30–46. London: Routledge.

Tomlinson, B. 2012. Materials development for language learning and teaching. *Language Teaching* 45, no. 2 (April): 143–179. https://doi.org/10.1017/S0261444811000528

Tomlinson, B. 2022. The discipline of materials development. In *The Routledge handbook of materials development for language teaching*, ed. J. Norton and H. Buchanan, 3–16. London: Routledge.

Uitto, A., and S. Saloranta. 2017. Subject teachers as educators for sustainability: A survey study. *Education Sciences* 7, no. 1 (January 4): 8. https://doi.org/10.3390/educsci7010008

Valnes Quammen, S., and O. Afitska. Forthcoming. Incorporating sustainability into American postsecondary modern foreign language classrooms: Exploring current practice and identifying future directions. *Manuscript available upon request from the authors.*

van Lier, L. 2004. *The ecology and semiotics of language learning: A sociocultural perspective*. Boston: Kluwer Academic Publishers.

Vitta, J.P. 2023. The functions and features of ELT textbooks and textbook analysis: A concise review. *RELC Journal* 54, no. 3 (December): 856–863.

Wagner, E. 2015. Survey research. In *Research methods in applied linguistics: A practical resource*, eds. B. Paltridge and A. Phakiti, 83–99. London: Bloomsbury Publishing.

Wagner, M., F. Cardetti, and M. Byram. 2018. The humble linguist: interdisciplinary perspectives on teaching intercultural citizenship. In *The talking species: Perspectives on the evolutionary, neuronal and cultural foundations of language*, ed. E.M. Luef, 423–447. Uni-Press Graz Verlag Graz.

Waltner, E.-M., K. Scharenberg, C. Hörsch, and W. Rieß. 2020. What teachers think and know about Education for Sustainable Development and how they implement it in class. *Sustainability* 12, no. 4 (February 24): 1690. https://doi.org/10.3390/su12041690

Wiek, A., L. Withycombe, and C.L. Redman. 2011. Key competencies in sustainability: A reference framework for academic program development. *Sustainability Science* 6 (2) (July): 203–218. https://doi.org/10.1007/s11625-011-0132-6

Wiek, A., M. Bernstein, R. Foley, M. Cohen, N. Forrest, C. Kuzdas, B. Kay, and L. Withycombe Keeler. 2015. Operationalising competencies in higher education for sustainable development. In *Handbook of higher education for sustainable development*, ed. M. Barth, G. Michelsen, M. Rieckmann, and I. Thomas, 241–260. London: Routledge.

9

The Social Aspect of Sustainable Development in Language Teaching and Learning—Pre-service Language Teachers' Perceptions of Multilingualism

Leena Maria Heikkola, Minna Maijala, Maarit Mutta, and Päivi Laine

Introduction

Tackling climate change is the ultimate goal of the United Nations' 2030 Agenda for Sustainable Development (United Nations 2015). The 2030 Agenda includes 17 Sustainable Development Goals (SDGs), which form an urgent call for action. All of 'the SDGs interface with

L. M. Heikkola (✉)
The Institute for Languages and Culture, The Arctic University of Norway, Tromsø, Norway
e-mail: leena.m.heikkola@uit.no

M. Maijala · M. Mutta · P. Laine
School of Languages and Translation Studies, University of Turku, Turku, Finland
e-mail: minmai@utu.fi

M. Mutta
e-mail: maamut@utu.fi

P. Laine
e-mail: paipel@utu.fi

M. Maijala et al. (eds.), *Transformative Language Teaching for Sustainability*,
https://doi.org/10.1007/978-3-031-85493-4_9

language, either as a substantive element of the goal itself (language as a goal) or as a means of communication, dialogue, response, and implementation (language as a tool)' (Language and the UN 2016). Thus, multilingualism is key to achieving the United Nations' 17 SDGs (see Stein-Smith 2016). Research on multilingualism has increased in recent years (e.g., Chik and Melo-Pfeifer 2023), but links between multilingualism and the achievement of the SDGs have not been fully explored.

In Finland, the context of the study, the current core curriculum for basic education also stresses the importance of strengthening the role of basic education in building a sustainable future (EDUFI 2014). The Finnish national core curriculum for basic education (EDUFI 2014) is quite progressive (Zilliacus et al. 2017) in its promotion of language awareness (van Lier 2010) and support for multilingualism. The curriculum reform (EDUFI 2014), which was a response to the growing number of migrant-background children in Finnish schools, has brought multilingualism to the forefront of all teaching, not just language teaching (see also Repo 2023; Zilliacus et al. 2017).

Although Finland is officially a bilingual country, the proportion of migrant-background inhabitants in Finland is relatively low compared to many other European countries (10%, Statistics Finland 2024). In the core curriculum, multilingualism is considered an essential resource for students' learning, and language awareness, linguistic identity, linguistic and cultural diversity, and learners' linguistic repertoire(s) are included in the core content of various subjects (EDUFI 2014). Thus, the core curriculum requires sustainability issues to be included in all subjects, including language teaching.

However, policy alone is not enough to support multilingualism in schools—it is teachers who implement language policies in the classroom (Hélot & Laoire 2011) and promote students' multilingualism (Haukås 2016). Still, educational policies and curricula influence how teachers engage with multilingualism and multilingual students (e.g., Tarnanen and Palviainen 2018). Studies have shown that in- and pre-service teachers in Finland have positive attitudes towards multilingualism but feel insecure about how to provide concrete support for multilingual students (Alisaari et al. 2019; Heikkola et al. 2022; for a global context, see Chik and Melo-Pfeifer 2023). Some of the reasons teachers have given for not using multilingual practices include a lack of suitable teaching materials, a lack of teacher training, teachers' personal linguistic repertoire, a lack of time, a lack of control, and inflexible school structures (Alisaari et al. 2019; Chik and Melo-Pfeifer 2023; Duarte and Günther-van der Meij 2022; Heikkola et al. forthcoming; Melo-Pfeifer 2021). As multilingualism is crucial in attaining the 17 SDGs presented in the Agenda 2030 by the United Nations, this study examines how Finnish pre-service language teachers understand multilingualism and its implementation in language classrooms.

Section 'The role of multilingual individuals and societies in promoting sustainable development' outlines different definitions of multilingualism based on the existing literature. Section 'Multilingualism in the classroom' discusses previous studies on multilingualism focusing on teachers' and learners' attitudes towards multilingualism and introduces the context of the study and our research questions. In the section 'Study design', the data collection instrument, participants, and methodology of the study are presented. In the 'Findings' section, the results are reported. Finally, in the 'Discussion' section, the findings are considered in relation to the literature and their implications for language teaching.

The Role of Multilingual Individuals and Societies in Promoting Sustainable Development

The European Commission (2007) defines multilingualism as 'the ability of societies, institutions, groups, and individuals to engage, on a regular basis, with more than one language in their day-to-day lives' (p. 6). Thus, multilingualism can be a characteristic of an individual or society. It is typical for members of multilingual society to speak more languages than members of monolingual society (Cenoz 2013). Often, plurilingualism is used as a synonym for multilingualism, although the Council of Europe (2007, p. 8) differentiates between the concepts, defining plurilingualism as a 'repertoire of varieties of language which many individuals use', whereas multilingualism is 'the presence in a geographical area, large or small, of more than one 'variety of language' i.e., the mode of speaking of a social group' (see also Cenoz 2013; Kalaja and Pitkänen-Huhta 2020). In this study, we define multilingualism 'as the generic term for anyone who speaks more than one language' (Ayrez-Bennet and Fisher 2022, p. 3); it includes both societal and individual characteristics, which are inseparable.

Skutnabb-Kangas (1981) defined individual multilingualism based on four different criteria: (1) origin, meaning a person is multilingual if they have spoken two or more languages since birth (also called simultaneous multilingualism, in contrast to sequential multilingualism; Silva-Corvalán 2018); (2) competence, meaning a person is multilingual if they are competent in two or more languages; (3) function, meaning a person is multilingual if they use two or more languages regularly in their everyday life; and (4) attitude/identity, meaning a person is multilingual if they identify as multilingual. At the societal level, multilingualism can be seen as additive (a language is added to a person's linguistic repertoire) or subtractive (a new language replaces a first language [L1]) (Cenoz 2013).

From an individual perspective, the use of language is closely related to one's identity (Edwards 2009/2012). Different languages can be used in different contexts (e.g., family, friends, school, and work), creating a multilingual identity (Ayrez-Bennet and Fisher 2022). Identity is constructed in social interaction; it is a life-long and dynamic process (see e.g., Evans 2018). Shared reflection on multilingual repertoires supports identity development, as it increases sociolinguistic awareness, facilitating 'practices of personalised reflection on one's own multilingual identity' (Gayton and Fisher 2022, p. 300). Valuing linguistic diversity and multilingualism in individuals, and thus in societies, contributes to democracy, social cohesion, and social justice (e.g., Chik et al. 2019). All languages are equally important in this respect (Piller 2016).

According to Bauman (2004), a person's identity contains some solid and some liquid parts; people have plural identities that can change in different situations. However, one cannot always freely choose one's identity—others' perceptions of us impact the way we see ourselves (Dervin and Jackson 2018). A multilingual person's self-perception might differ from that of others, for instance, in a classroom situation. However, multiculturalism in public spheres (e.g., school environments) can promote equality and social justice in terms of different cultural, racial, and religious groups; In order to achieve this, culturally responsive teaching is necessary (Gay 2018).

Multilingualism is particularly linked to SDG 4, Quality Education, through UNESCO guidelines and publications (e.g., Hult 2023). Promoting multilingualism is a contribution to the social cohesion and well-being of a society (e.g., Turner 2019). In sum, multilingualism enriches language teaching by fostering an inclusive learning environment where different linguistic backgrounds are valued. This not only improves cognitive and academic outcomes for students but also promotes linguistic diversity in society, ensuring the preservation and appreciation of different languages and cultures.

Multilingualism in the Classroom

Teachers' Beliefs about Multilingualism

In previous studies on pre-service teachers' beliefs about multilingualism in schools, the role of multilingualism has mostly been examined in the context of linguistically diverse classrooms. Accordingly, the concept of multilingualism is often linked to promoting linguistically responsive teaching (Alisaari et al. 2019; Fischer and Lahmann 2020; Heikkola et al. 2022; Repo 2020). In schools, a monolingual approach is still dominant (Alisaari et al. 2019); language subjects are taught separately, using as much target language as possible (see also Cummins 2007), and learners' language skills are evaluated in each individual language, disregarding multilingual linguistic repertoires (Shohamy 2011).

Multilingualism in today's classrooms should encompass each student's linguistic repertoires. In language-aware schools, multilingual, learner-centred pedagogies are used to develop learners' multiple languages (Cenoz and Gorter 2022). However, multilingualism is not always discussed positively, and, for example, migrants' multilingualism is often linked to societal problems (e.g., Haukås 2022).

Multilingual pedagogies are 'a pedagogical tool in multilingual education' (Cenoz and Gorter 2017, p. 309). The term *pedagogical translanguaging* encompasses instruction that allows learners to use their whole linguistic repertoire (e.g., Creese and Backledge 2010; Repo 2023); such instruction might be pre-planned, supporting the co-construction of knowledge among peers with the same linguistic repertoires, or spontaneous, giving opportunities for clarifications (see also Symons 2021). Pedagogical translanguaging is key to increasing affordances to multilingualism in classrooms; however, even if teachers are aware of linguistically responsive teaching, they may not be prepared to support linguistic diversity, and their thinking and actions may be contradictory (Repo 2023).

In the European context, there are studies available on teachers' attitudes towards multilingualism and multilingual learners. Schroedler and others (2023) examined German pre-service teachers' ($N = 259$) beliefs about multilingualism and language use; an unusually high number (62%) of participants were of migrant background and spoke languages other than German. The survey was conducted before and after completing a compulsory module on multilingualism using an instrument that captured the participants' beliefs about multilingual language use at home and in the classroom and about teachers' responsibility for classroom language support. The quantitative analysis was complemented by interviews investigating the participants' backgrounds. The findings showed significant improvement between the pre- and post-tests, indicating that beliefs regarding multilingualism and multilingual learners can develop even over a single semester. According to the findings, female participants were more likely to have positive beliefs about multilingualism than their male peers, and higher levels of education predicted more positive beliefs.

Young (2014) analysed interviews with head teachers ($N = 46$) in the Strasbourg region, revealing linguistic hierarchies and separate spaces for different languages. The majority of the participants understood multilingualism as bilingual practices in the regional languages of German and French or as learning German or English at school. Heritage languages were seldom mentioned. In the Netherlands, Bezcioglu-Göktolga and Yagmur (2018) examined the impact of Dutch classroom teachers on

Turkish immigrant families' language policies. Based on observations of 20 families and interviews with 35 parents and five teachers, the findings showed conflicts between parents and teachers. The parents doubted the teachers' professional advice, while the teachers were concerned about the parents' role in their children's schooling and their use of a heritage language at home. The teachers advised the parents to speak Dutch at home, but the parents wanted their children to maintain Turkish. Both parties wanted the children to be successful at school.

In a study involving 67 schools in Flanders, Strobbe and others (2017) examined the impact of schools' linguistic compositions on teachers' perceptions of multilingualism; faculty in more linguistically diverse schools showed increased tolerance towards multilingualism. How much the language of instruction was used and teachers' control thereof also seemed to influence tolerance towards multilingualism; thus, teachers' understanding of multilingualism depended on the context. Similarly, Lundberg (2019) examined subject teachers' ($N = 67$) views of multilingualism and pedagogical practices in German-speaking Switzerland, finding that all teachers, regardless of the subject they taught, considered themselves capable of communicating in more than one language. However, only the language teachers regarded their own teaching as an important part of students' multilingualism and seemed better at tailoring language policy and curriculum to students' needs. Disciplinary traditions and regional context also influenced understandings of multilingualism and related classroom practices.

De Angelis (2011) examined the beliefs of secondary school teachers ($N = 176$) in Italy ($n = 103$), Austria ($n = 42$), and Great Britain ($n = 31$) regarding prior language knowledge and its influence on teaching practices. The teachers shared similar views about multilingualism; most believed that knowing one language helped with learning additional languages, but they did not link language knowledge to better academic outcomes. The teachers held misconceptions that (home) languages can cause difficulties in or delay learning other languages. However, some teachers—those from Italy, in particular—seemed receptive to students' home languages and cultures.

Alisaari and others (2019) studied Finnish primary school classroom teachers', secondary and upper-secondary subject teachers', and special education teachers' beliefs about multilingualism and teaching multilingual learners (*N* = 820). The teachers' beliefs towards multilingualism were mainly positive; however, they maintained monolingual practices and did not recognise learners' home languages as a resource for learning. In another study using the same dataset, Alisaari and others (2021) examined Finnish teachers' (*N* = 691) beliefs about students' use of home languages in learning assignments and teachers' preparedness to implement linguistic awareness. Although the teachers' beliefs were mainly positive, only 7% encouraged home language use; 20% prevented it in the classroom, and 39% required students to present in the language of instruction. Repo (2020) investigated narratives of encountering multilingual learners using semi-structured group interviews with lower secondary school teachers (*n* = 16) working in linguistically diverse schools in Finland. The findings revealed contradictory discourses on encountering multilingual learners and supporting their learning, and problematic areas in teachers' expertise, such as their agency, acceptance, and willingness to change.

Palviainen and others (2016) examined bilingual preschool teachers (Finnish-Russian and Finnish-Swedish) in Finland and found that the teachers switched flexibly between languages when addressing children's needs. In a recent Finnish study focusing on extra-curricular heritage language courses for migrant children, Ansó Ros et al. (2021) investigated the impact of heritage teachers on students' multilingualism. The findings indicated that heritage language teachers support multilingual identities, show an understanding of students' multilingualism, and play an important role in supporting heritage language learners.

Suuriniemi and others (2021) studied teachers' attitudes towards multilingualism in 824 schools in Finland with 2864 teachers. Based on a semantic differential analysis of the teachers' responses, the teachers were divided into three groups: positive, reflective, and cautious. Positive teachers (18%) considered multilingualism as something that enriches and inspires the school environment and should thus be maintained, whereas reflective teachers (34%) saw multilingualism as a positive phenomenon but described it with negative adjectives, such as 'problematic' and 'cumbersome'. The biggest group (48%), cautious teachers, described multilingualism more negatively. Subject teachers were more cautious than primary school teachers, while teachers working in Swedish-speaking schools or in schools recognising linguistic diversity were the most positive.

There are also some studies looking at pre-service teachers' beliefs on multilingualism in schools. Portolés and Martí (2020) examined pre-service teachers' ($N = 121$), including pre-service language teachers', beliefs about learning and teaching English in a Spanish community. Their beliefs seemed to depend on individual factors, particularly the teacher trainees' linguistic backgrounds. Although most beliefs were monolingually oriented, respondents whose L1 was Catalan promoted multilingualism and minority languages more than those whose L1 was Spanish. Beliefs about multilingual instructional practices stemming from mass media seemed more difficult to change than those linked to academic or research-based principles. In the Nordic context, Iversen (2020) studied pre-service teachers' ($n = 24$) engagement with multilingualism during their field placement. The study, which focused on translanguaging practices, found that the participants were willing to support multilingual students but hesitant to utilise students' linguistic resources in their classroom practices. For instance, the participants reported that they had not prepared lessons with consideration for translanguaging practices.

Learners' Attitudes Towards Multilingualism

Students are rarely invited to reflect on their own multilingualism. Reasons for such a monolingual approach may include teachers' lack of knowledge (Haukås 2016) or teachers' worry about students not learning the majority language (Alisaari et al. 2022). Haukås and others (2022) examined Norwegian secondary school students' ($N = 593$) beliefs about the benefits of multilingualism. The benefits related to language learning were seen as more positive than those related to general cognitive advantages, and students with multilingual friends, who had lived abroad or had a migrant background, had significantly more positive beliefs about multilingualism. No significant relationship was detected between students' beliefs and the number of languages learned in school.

Haukås (2022) also provided insight into students' self-perceptions of multilingualism in Norwegian lower secondary schools. Altogether, 116 students aged 13–15 years were asked to define multilingualism, and their answers were analysed qualitatively. The findings showed that the students did not necessarily share scholars' and educational documents' positive and self-inclusive views of multilingualism but defined it as the knowledge or use of multiple languages. Some believed that competence in at least two languages was required to be considered multilingual, while others did not see themselves as multilingual or were unsure whether they were multilingual. Only a small number of participants linked multilingualism to migration or growing up with two languages.

Context of the Study

Finland, the context of this study, is a multilingual country. The national languages are Finnish and Swedish; thus, persons with Finnish or Swedish as their L1 are entitled to service in their L1 when dealing with national authorities (Constitution 731/1999, §17). Furthermore, the Constitution of Finland guarantees the linguistic rights of persons speaking Sámi, Romani, and sign language (Constitution 731/1999, §17). However, the Basic Education Act (628/1998) states that the language of instruction in schools is either Finnish or Swedish, so bilingual schools are not legally possible. There are some exceptions to this, such as immersion schools, in which classes are offered in two languages, often Swedish and Finnish (Peltoniemi and Bergroth 2022).

Everyone residing in Finland has the right to maintain and develop their language and culture (Constitution of Finland 731/1999, §6, §16, §17; see, e.g., Holm and Londen 2010). In basic education, schools are encouraged and subsidised to organise teaching in students' L1s (if other than Finnish or Swedish; EDUFI 2023). However, in practice, such teaching is only offered to approximately 38% of students with an L1 other than Finnish or Swedish (Tainio et al. 2019; Vipunen 2022).

The current Finnish core curriculum for basic education (EDUFI 2014) emphasises the importance of multiple linguistic and cultural identities and requires supporting students' multilingualism by allowing and promoting the use of all their linguistic resources, especially in language subjects. Thus, multilingual practices and linguistic awareness are a requirement for all teachers in Finnish basic education (EDUFI 2014). This includes understanding the role of language in all learning and teaching and using all of the students' linguistic resources—and thus also supporting multilingualism—for learning. This requirement is especially important for language teachers in basic education and an important factor in pre-service language teachers' teacher training.

In Finland, pre-service language teachers complete a year-long teacher training in addition to their subject studies (major subject = 120 European Credit Transfer System [ECTS], minor subject = 60 ECTS; 1 ECTS equals 25–27 hours of work). The teacher training is generally integrated into master's programmes, but some students participate in the teacher training during their bachelor's degree. In teacher training, students learn practical skills by working in a real classroom supervised by mentoring teachers (see e.g., Pollari et al. 2018).

Aim and Research Questions

Although multilingual policies have entered societies and curricula, everyday school practices remain mainly monolingual. Pre-service teachers favour a multilingual approach but hesitate to include multilingual practices in their classrooms. Language learners (Haukås 2022) and teachers often have strict criteria for who is considered multilingual. Personal (Portolés and Martí 2020) and regional backgrounds and curricula (Lundberg 2019) seem to influence teachers' perceptions of multilingualism.

In promoting multilingualism in schools, teachers' and pre-service teachers' beliefs about and understandings of multilingualism are crucial (e.g., Chik and Melo-Pfeifer 2023)—without teachers' commitment, multilingual pedagogies will not be enacted (e.g., Ruohotie-Lyhty and Moate 2016). Thus, more research is required to understand how teachers can support multilingualism through classroom practices and how teacher education should be structured to reflect on multilingual diversity and its realisation in the classroom. As multilingualism and language and communication are a vital part of attaining especially the social aspects of the SDGs outlined in the Agenda 2030, teachers need an understanding of how to promote them in education.

The aim of this study was to investigate pre-service language teachers' understanding of multilingualism as a part of sustainable development. The focus of the study is thus on the social aspects of sustainable development. The following research questions (RQs) guided the study:

(RQ1) How do pre-service language teachers define multilingualism?
(RQ2) What kinds of actions constitute multilingualism or a multilingual person according to the pre-service teachers?
(RQ3) How is multilingual language use described by the pre-service teachers?

Study design

In this section, we present the data collection instrument and the participants. We also introduce the type of data analysis used in the study.

Instrument

The data were collected via an online survey investigating respondents' understandings of linguistically responsive practices, including multilingualism and the importance of L1s (Heikkola et al. 2021; Heikkola et al. 2022). The survey was based on the framework proposed by Lucas and Villegas (2013). The survey, which was an abridged version of a longer Finnish survey (see Alisaari et al. 2019) based on an English-language version (Milbourn et al. 2017), included 35 Likert scale statements and five open-ended questions. Responses to the following two-part open-ended question were analysed for this study: 'What is multilingualism? Who is multilingual?'.

Participants

The survey was introduced during a course on language and learning in the autumn of 2020. The course was a requirement for a master's degree in language learning and teaching. Authors 2, 3, and 4 were the teachers of the course at the time the survey was administered. The survey, which included a separate privacy notice, was presented after a distance lecture at the end of the course.

There were 82 students in the course, 52 (63%) of whom completed the survey. The participants were taking the one-year teacher education programme as a part of their MA degree. Our focus was on pre-service language teachers; thus, four students with majors other than language were removed from the analysis. The responses of 48 students were included in the analysis. The students responded anonymously after giving written consent to participate in the study. The students' background information is presented in Table 9.1.

Data Analysis

Data-driven content analysis was used to analyse the responses (see e.g., Krippendorf 2012). The responses were read by all the authors and categorised into the following: (1) definition of multilingualism as societal or individual, (2) multilingualism as actions, and (3) use of language in relation to multilingualism. Consensus for the categorisation of the responses was reached through discussions among the authors. To support our analysis, examples of responses are presented in the findings. The examples are followed by a running identification number assigned to the students. The original responses were given in Finnish and translated into English by the authors.

Table 9.1 Students' background information

Background factors	Students ($N = 48$)
Age	Mean = 25 (min. 22; max. 37)
L1	Finnish ($n = 47$) Finnish and Estonian ($n = 1$)
Lived abroad	Yes ($n = 25$) No ($n = 20$)
Schooling in language(s) other than L1	Yes ($n = 28$) No ($n = 13$)
Major	English ($n = 17$) Finnish/ Finno-Ugric Languages ($n = 12$) Scandinavian Languages ($n = 5$) Spanish ($n = 5$) French ($n = 3$) German ($n = 3$) Italian ($n = 3$)
Year of beginning studies	2020 ($n = 1$) 2019 ($n = 2$) 2018 ($n = 4$) 2017 ($n = 15$) 2016 ($n = 16$) 2015 or before ($n = 10$)
Academic year of teacher training	2021 or after ($n = 9$) 2020 ($n = 6$) 2019 ($n = 17$) 2018 ($n = 2$) Other ($n = 3$) No studies yet ($n = 11$)
Gender	Female ($n = 37$) Male ($n = 11$) Other ($n = 0$)

Findings

Here, we present the participants' definitions of multilingualism, the actions they attached to multilingualism, and outline their understandings of language used to describe multilingualism.

Who Is Multilingual?

Multilingualism was defined as either an individual or a community characteristic. All except one participant ($n = 47$) defined multilingualism as an individual characteristic (Examples 1–2).

(1) A person speaks several languages and is fluent in them. (ID01)
(2) A multilingual person regularly speaks at least two languages in everyday life. (ID03)

27% of the participants ($n = 13$) defined multilingualism as a community characteristic. In these responses, many defined multilingualism in terms of the Finnish context (Examples 3–4).

(3) Almost all Finns can speak English or Swedish in addition to Finnish, so I myself am multilingual. (ID04)
(4) If multilingualism is only linked to language skills, then everyone in Finland is multilingual, because here language teaching is at a high level, and almost everyone knows English. (ID20)

In most cases (12/13), the participants who defined multilingualism as a feature of a community indicated that an individual can also be multilingual (Examples 3, 5, and 6).

(5) A community or individual who speaks two or more languages. (ID22)
(6) Everyone is multilingual. Multilingual does not just mean someone who speaks several languages; everyone is multilingual. Why? Because different languages are often used in different situations, even if the language itself (e.g., Finnish) is the same. The language used at work and during leisure time is different. Of course, multilingualism is also the ability to speak several languages. (ID37)

The participants who defined multilingualism as a community feature often also defined multilingualism as control of different registers within the same language (Example 7).

(7) The different registers of language used in different situations can also be considered multilingualism, e.g., the languages of different subjects at school, everyday language, class language, official/formal language … (ID31)

What Does a Multilingual Do?

The respondents used a variety of verbs to describe actions related to multilingualism. Over half ($n = 27$) of the respondents described a multilingual person as someone who knows or has command of at least two languages (Examples 8–11).

(8) A multilingual person is anyone who speaks more than one language. (ID23)
(9) A multilingual person is someone who knows and uses more than one language. (ID26)
(10) Typically, a multilingual person often means a person who is fluent in several languages. (ID15)
(11) Being multilingual means being able to use the language fluently and being able to have a conversation. (ID38)

Two respondents believed that a multilingual person does not need a perfect or native-like command of all their languages (Examples 12–13).

(12) I think multilingualism means speaking/understanding more than one language. You don't have to know it at a native proficiency level. (ID27)
(13) A multilingual person does not need to be able to speak dozens of different languages but to know something about these languages and related cultures, for example, in order to communicate in a situation. (ID32)

In addition to knowing a language, many participants ($n = 20$) emphasised that a multilingual person (or community) uses more than one language (Examples 14–15).

(14) The person speaks several languages and is fluent in them. (ID01)
(15) Multilingualism is a matter of identity, but [...] a multilingual person is someone who actively uses more than one language on a daily basis. (ID20)

Many respondents (n = 20) also viewed speaking or producing in more than one language as a sign of multilingualism, whether in addition to or as a mother tongue (Examples 16–17).

(16) Multilingualism means being fluent in at least one language, in addition to your mother tongue. (ID42)
(17) Multilingualism means having people in your community who speak different languages as their mother tongue. (ID50)

Three participants stated that multilingualism was the ability to communicate or function in more than one language (Example 18).

(18) A multilingual person can function in everyday situations using appropriate language skills and languages. (ID28)

The responses revealed two types of multilinguals: people who had studied one or more languages (n = 8; Examples 19–20) and those who spoke another language in addition to the majority language (n = 18; Examples 21–23).

(19) Multilingualism is an asset and is also pursued at school. Learning more than one language has nothing but benefits. (ID36)
(20) You are multilingual if you have learned to speak a language at school or while working abroad. (ID38)
(21) I consider a multilingual student to be one who knows more than one language, with at least two having been learned outside school. (ID25)
(22) A multilingual person is, for example, someone whose mother tongue is a particular language and who lives in a country where the dominant language is different. (ID21)

(23) A multilingual person is a person who can speak several languages or is bilingual; nowadays, the term is often used for a person who speaks a minority language but studies [...] in the main language, for example, immigrants. (ID50)

Some students differentiated between having 'full' command of a language, distinguishing between mother tongues and learner languages (Examples 24–25) versus having learned many languages in school (Examples 26–27).

(24) Multilingualism means speaking one or more languages. A multilingual person is someone who has used or uses two or more languages in parallel. However, this does not include, for example, having to make business calls in a language other than one's mother tongue. (ID35)
(25) I would not consider a Finnish-speaking pupil who has learned Swedish and English in a school environment to be multilingual. (ID25)
(26) Being multilingual means being able to speak several languages fluently, excluding, for example, knowing a few words/phrases in a language. (ID16)
(27) They may have one mother tongue but learned several languages since then. (ID44)

Although many participants stated that multilingualism meant speaking or being able to use languages, only five wrote that understanding more than one language was part of multilingualism (Examples 28–29).

(28) I think multilingualism means speaking/understanding more than one language. You don't have to know it at a native speaker's proficiency level. (ID27)
(29) A person who understands more than one language. (ID30)

There were also mentions of identifying as multilingual and reading and writing in more than one language.

How Is Multilingual Language Use Described?

Several modifiers were used to describe multilingual language use. Multilingualism was most often seen as the frequent use of many ($n = 12$; Example 30) or different languages ($n = 10$; Example 31).

(30) A multilingual person is someone who speaks several languages. (ID48)
(31) A person who works in many different languages in everyday life. (ID08)

Participants also described languages being used fluently ($n = 10$; Example 32), regularly ($n = 3$) and in everyday life ($n = 6$; Example 33), actively ($n = 3$; Example 34), and officially ($n = 2$; Example 35).

(32) A multilingual person is someone who speaks two foreign languages fluently in addition to their mother tongue. (ID05)
(33) A multilingual person regularly speaks at least two languages in everyday life. (ID03)
(34) Multilingualism can relate to a society or community in which several languages are actively used. (ID42)
(35) Multilingualism can mean a person speaking several languages or, for example, being officially bilingual or multilingual by law. (ID33)

There was also mention of using specific languages; using language skilfully, typically, or daily; and having learned language(s) naturally (Example 36).

(36) Fluent in two or more languages. Some definitions also specify that language learning must have taken place naturally before a certain age. (ID42)

Discussion

Most participants considered multilingualism to be an individual characteristic. This aligns with earlier studies arguing that multi- and bilingualism are often related to persons who acquire two or more languages from birth (origin criterion; Skutnabb-Kangas 1981), although only a few Norwegian secondary school students linked multilingualism to growing up with two or more languages (Haukås 2022). The participants who defined multilingualism as a community feature indicated that individuals can also be multilingual. Thus, some participants understood that societal and individual multilingualism are inseparable (Ayrez-Bennet and Fisher 2022). Those who viewed multilingualism as both a community and an individual characteristic also understood that the control of multiple registers within the same language counts as multilingualism.

There was a diverse range of perspectives regarding the actions and characteristics of multilingual individuals. More than half of the respondents described a multilingual person as someone who knows or commands at least two languages. This indicates the belief that multilingualism entails being proficient in multiple languages (competence criterion; Skutnabb-Kangas 1981). This is in line with a previous Norwegian study by Haukås (2022) looking at secondary school students, although some of these students did not apply this definition to themselves. However, not all participants required a fluent command of all languages (origin criterion; Skutnabb-Kangas 1981). Thus, the participants understood multilingualism as both simultaneous and sequential (Silva-Corvalán 2018).

Two-thirds of the participants emphasised that multilingualism involves the frequent use of multiple languages (function criterion; Skutnabb-Kangas 1981), highlighting the dynamic nature of multilingualism. In addition, almost half of the respondents associated multilingualism with speaking or producing more than one language, not just passively understanding languages. This indicates a belief that actively engaging in the use of multiple languages equates to multilingualism (on active–passive bilingualism, see Sharwood Smith 2017).

Within the pre-service teachers' responses, two types of multilinguals were identified: individuals who had studied one or more languages and individuals who spoke another language in addition to the majority language. This suggests that multilingualism encompasses both formal language study, as well as belonging to bilingual or multilingual communities (see e.g., Silva-Corvalán 2018; Walton 2009). Thus, it seems that at least some understood the importance of language to one's identity (see e.g., Edwards 2009/2012) and that languages can be used in different contexts (for example, school and home, see e.g., Ayrez-Bennet and Fisher 2022). Some participants distinguished between having 'full' command of a language and having acquired knowledge of multiple languages through school, implying that proficiency levels in each language can vary (e.g., Kemp 2009), with some languages being learned more extensively through formal education (Walton 2009).

In addition to illustrative verbs (e.g., know, command, use, speak, and produce), the participants employed various adjectives and adverbs to describe language use connected to multilingualism, saying multilingualism consists of the use of many or different languages and multiple languages are used fluently, in everyday life, frequently, actively, and officially; and language use is native-like. Some participants also mentioned using languages skilfully, typically, daily, or naturally.

Our results show that pre-service language teachers have a strong developing understanding of different aspects of multilingualism. We can thus assume that they are at least somewhat equipped to promote multilingualism in their teaching—as required by the Finnish core curricula—which contributes to the social cohesion and well-being of society (Turner 2019). Promoting multilingualism can thus contribute to the achievement of the SDGs (Hult 2023)—particularly the social aspects of sustainable development. By educating prospective language teachers to promote multilingualism, they can improve the academic outcomes for all their students and foster inclusive language learning environments where different linguistic backgrounds are valued and made visible. However, when addressing sustainability issues in multi-ethnic contexts in the Global South, it is important to consider the challenges encountered, e.g., by local Indigenous groups (see Chapter 10).

Conclusion

The majority of pre-service language teachers perceived multilingualism as the ability to know or command at least two languages. However, there was recognition that fluency in all languages is not required. The use of multiple languages and active engagement in their production were considered important factors, and there was acknowledgement of different types of multilingualism within communities. Overall, the participants seemed aware of the many forms of multilingualism. An understanding of the different facets of multilingualism equips teachers to support students using multilingual pedagogies and thus promote the sustainability goals set by the UN. However, more research is needed to examine the actual multilingual practices used in the classroom.

References

Alisaari, J., L.M. Heikkola, N. Commins, and E.O. Acquah. 2019. Monolingual ideologies confronting multilingual realities. Finnish teachers' beliefs about linguistic diversity. *Teaching and Teacher Education* 80 (April): 48–58. https://doi.org/10.1016/j.tate.2019.01.003

Alisaari, J., T.-R. Hurme, L.M. Heikkola, and S. Routarinne. 2021. Finnish teacher's beliefs about students' home language use. *Critical Multilingualism Studies* 9 (1): 46–76. Available from: https://cms.arizona.edu/index.php/multilingual/article/view/223/317. Accessed October 4, 2024.

Ansó Ros, J., M. Maijala, and N. Valkamo. 2024. The role of the teacher in heritage language maintenance courses in Finland. *Journal of Multilingual and Multicultural Development* 45, no. 2 (February 7): 522–535. https://doi.org/10.1080/01434632.2021.1906692.

Ayrez-Bennet, W., and L. Fisher. 2022. *Multilingualism and identity. Interdisciplinary perspectives.* Cambridge: Cambridge University Press.

Bauman, Z. 2004. *Identity*. Cambridge: Polity.

Bezcioglu-Göktolga, I., and K. Yagmur. 2018. The impact of Dutch teachers on family language policy of Turkish immigrant parents. *Language, Culture and Curriculum* 31, no. 3 (September 2): 220–234. https://doi.org/10.1080/07908318.2018.1504392.

Cenoz, J. 2013. Defining multilingualism. *Annual Review of Applied Linguistics* 33 (March): 3–18. https://doi.org/10.1017/S026719051300007X.

Cenoz, J., and D. Gorter. 2017. Translanguaging as a pedagogical tool in multilingual education. In *Language Awareness and Multilingualism*, Ed. Cenoz J., D. Gorter, and S. May, 309–321. Cham: Springer International Publishing. https://doi.org/10.1007/978-3-319-02240-6_20.

Cenoz, J., and D. Gorter. 2022. Pedagogical translanguaging in content and language integrated learning. *Journal of Multilingual Theories and Practices* 3.1 (May 10): 7–26. https://doi.org/10.1558/jmtp.21812.

Chik, A., J. Forrest, and F. Siciliano. 2019. Language diversity in Sydney: At home and in public. In *Multilingual Sydney*, ed. A. Chik, P. Benson, and R. Mononey, 26–39. London, New York: Routledge.

Chik, A., and S. Melo-Pfeifer. 2023. Discussion of the special issue. Beliefs about multilingualism: Is a methodological and epistemological change necessary? *International Journal of Multilingualism* 20, no. 1 (January 2): 86–96. https://doi.org/10.1080/14790718.2023.2172174.

Constitution of Finland. 731/1999, §6, §16, §17.

Council of Europe. 2007. *From linguistic diversity to plurilingual education: Guide for the development of language education policies in Europe*. Available from: https://rm.coe.int/16802fc1c4. Accessed October 4, 2024.

Creese, A., and A. Blackledge. 2010. Translanguaging in the bilingual classroom: A pedagogy for learning and teaching? *The Modern Language Journal* 94 (1) (March): 103–115. https://doi.org/10.1111/j.1540-4781.2009.00986.x

Cummins, J. 2007. Rethinking monolingual instructional strategies in multilingual classrooms. *Canadian Journal of Applied Linguistics* 10 (2): 221–240. Available from: https://journals.lib.unb.ca/index.php/CJAL/article/view/19743. Accessed October 4, 2024.

De Angelis, G. 2011. Teachers' beliefs about the role of prior language knowledge in learning and how these influence teaching practices. *International Journal of Multilingualism* 8, no. 3 (August): 216–234. https://doi.org/10.1080/14790718.2011.560669.

Dervin, F., and J. Jackson. 2018. Language, identity and interculturality. In *Interculturality in international education*, ed. J. Jackson, 63–81. New York: Routledge.

Duarte, J., and M. Günther-van Der Meij. 2022. 'Just accept each other, while the rest of the world doesn't'–Teachers' reflections on multilingual education. *Language and Education* 36, no. 5 (September 3): 451–466. https://doi.org/10.1080/09500782.2022.2098678.

EDUFI. 2014. *National core curriculum for basic education.* Available from: https://www.oph.fi/sites/default/files/documents/perusopetuksen_opetussuunnitelman_perusteet_2014.pdf. Accessed August 31, 2024.

EDUFI. 2023. *Objectives, contents and assessment of learning and competence in the mother tongue of pupils in supplementary basic education.* Available from: https://www.oph.fi/sites/default/files/documents/LIITE%203%20PERUSOPETUSTA%20T%C3%84YDENT%C3%84V%C3%84N%20OPPILAAN%20OMAN%20%C3%84IDINKIELEN%20OPETUKSEN%20TAVOITTEET_2.pdf. Accessed October 4, 2024.

Edwards, J. 2009/2012. *Language and identity. Key topics in sociolinguistics.* Cambridge: Cambridge University Press.

European Commission. 2007. *Final report: High level group on multilingualism.* Available from: https://op.europa.eu/en/publication-detail/-/publication/b0a1339f-f181-4de5-abd3-130180f177c7. Accessed October 4, 2024.

Evans, D. 2018. *Language, identity and symbolic culture.* London: Bloomsbury Academic.

Fischer, N., and C. Lahmann. 2020. Pre-service teachers' beliefs about multilingualism in school: An evaluation of a course concept for introducing linguistically responsive teaching. *Language Awareness* 29, no. 2 (April 2): 114–133. https://doi.org/10.1080/09658416.2020.1737706.

Gay, G. 2018. *Culturally responsive teaching: Theory, research, and practice*, 3rd ed. New York: Teachers College Press.

Gayton, A., and L. Fisher. 2022. Multilingual multilingualism and identity: Interdisciplinary perspectives, identity construction through participative reflective practice in the language classroom. In *Multilingualism and identity*, ed. W. Ayres-Bennett and L. Fisher, 281–298. Oxford: Oxford University Press.

Haukås, Å. 2016. Teachers' beliefs about multilingualism and a multilingual pedagogical approach. *International Journal of Multilingualism* 13, no. 1 (January 2): 1–18. https://doi.org/10.1080/14790718.2015.1041960.

Haukås, Å. 2022. Who are the multilinguals? Pupil's definitions, self-perceptions and public debate. In *Multilingualism and identity*, ed. W. Ayres-Bennett and L. Fisher, 281–298. Oxford: Oxford University Press.

Heikkola, L.M., J. Alisaari, H. Vigren, and N. Commins. 2022. Requirements meet reality: Finnish teachers' practices in linguistically diverse classrooms. *Journal of Language, Identity & Education* 23, no. 4 (July 3): 433–449. https://doi.org/10.1080/15348458.2021.1991801.

Heikkola, L.M., E. Repo, and N. Kekki. 2021. Kielen rooli oppimisessa—Tulevien aineenopettajien käsityksiä [The role of language in learning and teaching: Perceptions of subject teacher students]. In *Språkets function [Functions of language]*, ed. L.M. Heikkola, G. Paulsen, K. Wojciechowicz, and J. Rosenberg, 330–351. Åbo: Åbo Akademi Press.

Heikkola, L.M., E. Repo, and N. Kekki. (forthcoming 2024). Mapping pre-service subject teachers' preparedness for linguistically diverse classrooms in Finland. In *Teacher education for working in linguistically diverse classrooms: Nordic perspectives*, ed. J. Iversen, A.R. Warren, and B. Straszer. Berlin: Language Science Press.

Hélot, C., and M.Ó. Laoire, eds. 2011. *Language policy for the multilingual classroom: Pedagogy of the possible*. Bristol: Multilingual Matters.

Holm, G., and M. Londen. 2010. The discourse on multicultural education in Finland: Education for whom? *Intercultural Education* 21, no. 2 (April): 107–120. https://doi.org/10.1080/14675981003696222.

Hult, F.M. 2023. Sustainable multilingual education. In *Language and sustainable development*, ed. L.J. McEntee-Atalianis and H. Tonkin, 51–77. Cham: Springer.

Iversen, J.Y. 2020. Pre-service teachers' translanguaging during field placement in multilingual, mainstream classrooms in Norway. *Language and Education* 34, no. 1 (January 2): 51–65. https://doi.org/10.1080/09500782.2019.1682599.

Kalaja, P., and A. Pitkänen-Huhta. 2020. Raising awareness of multilingualism as lived—In the context of teaching English as a foreign language. *Language and Intercultural Communication* 20, no. 4 (July 3): 340–355. https://doi.org/10.1080/14708477.2020.1786918.

Kemp, C. 2009. Defining multilingualism. In *The exploration of multilingualism: Development of research on L3, multilingualism and multiple language acquisition.* [AILA Applied Linguistics Series 6], ed. L. Aronin, and B. Hufeisen, 11–26. Amsterdam/Philadelphia: John Benjamins. https://doi.org/10.1075/aals.6.02ch2.

Krippendorf, K. 2012. *Content analysis: An introduction to its methodology*, 3rd ed. Thousand Oaks: Sage Publications.

Language and the UN. 2016. *Symposium on language and the Sustainable Development Goals*. Available from: https://sites.google.com/site/languageandtheun/events/2016-symposium-conclusions. Accessed August 31, 2024.

Lucas, T., and A.M. Villegas. 2013. Preparing linguistically responsive teachers: Laying the foundation in preservice teacher education. *Theory Into Practice* 52, no. 2 (April): 98–109. https://doi.org/10.1080/00405841.2013.770327.

Lundberg, A. 2019. Teachers' viewpoints about an educational reform concerning multilingualism in German-speaking Switzerland. *Learning and Instruction* 64 (December): 101244. https://doi.org/10.1016/j.learninstruc.2019.101244.

Melo-Pfeifer, S. 2021. Exploiting foreign language student-teachers' visual language biographies to challenge the monolingual mind-set in foreign language education. *International Journal of Multilingualism* 18 (4): 601–618.

Milbourn, T., K.M. Viesca, and N. Leech. 2017. Measuring linguistically responsive teaching: First results. Paper presented at the American educational researchers' association annual meeting, San Antonio, TX, April 16, 2017.

Palviainen, Å., E. Protassova, K. Mård-Miettinen, and M. Schwartz. 2016. Two languages in the air: A cross-cultural comparison of preschool teachers' reflections on their flexible bilingual practices. *International Journal of Bilingual Education and Bilingualism* 19, no. 6 (November): 614–630. https://doi.org/10.1080/13670050.2016.1184615.

Peltoniemi, A., and M. Bergroth. 2022. Developing language-aware immersion teacher education: Identifying characteristics through a study of immersion teacher socialisation. *International Journal of Bilingual Education and Bilingualism* 25, no. 4 (April 21): 1324–1335. https://doi.org/10.1080/13670050.2020.1757613.

Piller, I. 2016. *Linguistic diversity and social justice: An introduction to applied sociolinguistics.* New York: Oxford University Press.

Pollari, P., O. Salo, and K. Koski. 2018. In teachers we trust—The Finnish way to teach and learn. *Inquiry in Education* 10 (1): article 4. Available from: https://files.eric.ed.gov/fulltext/EJ1180628.pdf. Accessed August 31, 2024.

Portolés, L., and O. Martí. 2020. Teachers' beliefs about multilingual pedagogies and the role of initial training. *International Journal of Multilingualism* 17 (2): 248–264. https://doi.org/10.1080/14790718.2018.1515206.

Repo, E. 2020. Discourses on encountering multilingual learners in Finnish Schools. *Linguistics and Education* 60 (December): 100864. https://doi.org/10.1016/j.linged.2020.100864.

Repo, E. 2023. Together towards language-aware schools. Perspectives on supporting increasing linguistic diversity. PhD Thesis, The University of Turku. Available from: https://urn.fi/URN:ISBN:978-951-29-9178-5. Accessed August 31, 2024.

Ruohotie-Lyhty, M., and J. Moate. 2016. Who and how? Preservice teachers as active agents developing professional identities. *Teaching and Teacher Education* 55 (April): 318–327. https://doi.org/10.1016/j.tate.2016.01.022.

Schroedler, T., H. Rosner-Blumenthal, and C. Böning. 2023. A mixed-methods approach to analysing interdependencies and predictors of pre-service teachers' beliefs about multilingualism. *International Journal of Multilingualism* 20, no. 1 (January 2): 11–30. https://doi.org/10.1080/14790718.2022.2125975.

Sharwood Smith, M. 2017. Active-passive bilingualism and the functional distance between L1 and L2 as viewed within one unifying theoretical framework. In *Psychology of bilingualism. The bilingual mind and brain book series.* ed. A. Ardila, A. Cieślicka, R. Heredia, and M. Rosselli. 157–188, Berlin: Springer.

Shohamy, E. 2011. Assessing multilingual competencies: Adopting construct valid assessment policies. *The Modern Language Journal* 95, no. 3 (September): 418–429. https://doi.org/10.1111/j.1540-4781.2011.01210.x.

Silva-Corvalán, C. 2018. Simultaneous bilingualism: Early developments, incomplete later outcomes? *International Journal of Bilingualism* 22, no. 5 (October): 497–512. https://doi.org/10.1177/1367006916652061.

Skutnabb-Kangas, T. 1981. *Bilingualism or not. The education of minorities.* Bristol: Multilingual Matters.

Stein-Smith, K. 2016. The role of multilingualism in effectively addressing global issues: The Sustainable Development Goals and beyond. *Theory and Practice in Language Studies* 6, no. 12 (December 1): 2254. https://doi.org/10.17507/tpls.0612.03.

Statistics Finland. 2024. *Immigrants in integration.* Available at: https://stat.fi/tup/maahanmuutto/index_en.html. Accessed March 2, 2025.

Strobbe, L., A. Van Der Wildt, P. Van Avermaet, K. Van Gorp, K. Van Den Branden, and M. Van Houtte. 2017. How school teams perceive and handle multilingualism: The impact of a school's pupil composition. *Teaching and Teacher Education* 64 (May): 93–104. https://doi.org/10.1016/j.tate.2017.01.023.

Suuriniemi, S.-M., M. Ahlholm, and V. Salonen. 2021. Opettajien käsitykset monikielisyydestä: heijastumia koulun kielipolitiikasta [Teachers' views on multilingualism: Reflections on the school's language policy]. In *Koulun monet kielet [Plurilingualism in the school*], AFinLA-e 13, ed. by M. Ahlholm, and P. Portaankorva-Koivisto, 44–69. https://doi.org/10.30660/afinla.100518

Symons, C. 2021. Instructional practices for scaffolding emergent bilinguals' comprehension of informational science texts. *Pedagogies: An International Journal* 16, no. 1 (January 2): 62–80. https://doi.org/10.1080/1554480X.2020.1738938.

Tainio, L., A. Kallioniemi, R. Hotulainen, M. Ahlholm, R. Ahtiainen, M. Asikainen, L. Avelin, I, Grym, J. Ikkala, M. Laine, N. Lankinen, K. Lehtola, E. Lindgren, I. Rämä, T. Sarkkinen, M. Tamm, E. Tuovila, and N. Virkkala. 2019. *Koulujen monet kielet ja uskonnot. Selvitys vähemmistöäidinkielten ja -uskontojen sekä suomi ja ruotsi toisena kielenä -opetuksen tilanteesta eri koulutusasteilla [The many languages and religions of the schools. Survey of the situation of minority mother tongues and religions and the teaching of Finnish and Swedish as a second language at different levels of education*]. The Prime Minister's Office. Available from: http://urn.fi/URN:ISBN:978-952-287-640-9. Accessed August 31, 2024.

Tarnanen, M., and Å. Palviainen. 2018. Finnish teachers as policy agents in a changing society. *Language and Education* 32, no. 5 (September 3): 428–443. https://doi.org/10.1080/09500782.2018.1490747.

Turner, M. 2019. *Multilingualism as a resource and a goal: Using and learning languages in mainstream schools*. Cham: Springer International Publishing.

United Nations. 2015. *The 17 Goals*. Available from: https://sdgs.un.org/goals. accessed August 31, 2024.

Lier, L.V. 2010. The ecology of language learning: Practice to theory, theory to practice. *Procedia—Social and Behavioral Sciences* 3: 2–6. https://doi.org/10.1016/j.sbspro.2010.07.005.

Vipunen. 2022. *Pupils in pre-primary education and basic education*. Available from: https://vipunen.fi/en-gb/basic/Pages/Oppilaat-ja-perusopetuksen-p%C3%A4%C3%A4tt%C3%A4neet.aspx. Accessed March 2, 2025.

Walton, A. 2009. Multilingualism and foreign language learning. In *Handbook of foreign language communication and learning*, ed. K. Knapp and B. Seidlhofer, 45–78. Berlin, New York: De Gruyter.

Young, A.S. 2014. Unpacking teachers' language ideologies: Attitudes, beliefs, and practiced language policies in schools in Alsace, France. *Language Awareness* 23, no. 1–2 (April 3): 157–171. https://doi.org/10.1080/09658416.2013.863902.

Zilliacus, H., B. Paulsrud, and G. Holm. 2017. Essentializing vs. non-essentializing students' cultural identities: Curricular discourses in Finland and Sweden. *Journal of Multicultural Discourses* 12, no. 2 (April 3): 166–180. https://doi.org/10.1080/17447143.2017.1311335.

10

Integrating Plurilingual Repertoires in a Multi-Ethnic School—Addressing Sustainability Issues in Amazonia

Santiago Sánchez Moreano

Introduction

This chapter discusses the specificities and challenges related to students' ethnicity, language, and socio-environmental context in terms of integrating plurilingual repertoires within a multi-ethnic Indigenous school in the periphery of Leticia—a Colombian pluriverse city with a high component of Indigenous populations (Echeverri Restrepo 2023) located along the triple border of Colombia, Brazil, and Peru in the heart of Amazonia. The use of several languages in education can lay the groundwork for linguistic citizenship (Williams and Stroud 2015) and social justice (Mohanty et al. 2009), as all languages and language varieties have a legitimate place in the world; furthermore, involving marginalised people in environmental conservation practices can lead to environmental justice (Brondízio et al. 2021).

S. Sánchez Moreano (✉)
French Institute of Research for Development (IRD), Montpellier, France
e-mail: santiago.sanchez-moreano@ird.fr

M. Maijala et al. (eds.), *Transformative Language Teaching for Sustainability*,
https://doi.org/10.1007/978-3-031-85493-4_10

The Colombian Amazonia is a region of high vulnerability due to the illegal exploitation of resources, the presence of illegal groups (de Rementería 2011), and a significant flow of national and international tourists (Acevedo 2011; Ochoa Zuluaga 2019). It represents 6.8% of the entire Amazon biome and is home to complex ecosystems and numerous ethnic groups, such as the Tikunas, Uitotos, Boras, Yucunas, and Cocamas. Most of these groups are transnational, since they live along and cross the borders on a daily basis and interact through family and social relationships, friendship and social activities, sociocultural manifestations, business relations, and economic interests (Gasché and Vela Mendoza 2011). For example, someone can live or work on the Colombian side, with relatives living or working on the Peruvian or Brazilian side, while maintaining their family or economic relations; someone can be born in one of these countries but grow up on the other side of the frontier.

The school under study is attended by 16 ethnic groups, and while this diversity represents a challenge for any inclusive curriculum, the school community faces structural social problems, such as discrimination, domestic violence, dysfunctional families, abandoned children, drug consumption, and high rates of school dropout (+20%). In this context, talking about sustainability and sustainability issues, such as deforestation, food supply systems, and eco-tourism, the latter being the pedagogical orientation of the school, is the least of the concerns of these students. Consequently, introducing Indigenous languages, for example, plurilingual repertoires, in the teaching practices of this school raises important challenges that could translate into opportunities for epistemic and environmental justice. In section 'Language diversity in education in Colombia', I briefly outline how the Colombian educational system has dealt with language and language diversity in the last three decades. Thereafter, I describe the specificities of the context in terms of social and linguistic constraints. This is followed by a brief description of the methods used in this study, after which I discuss the challenges and issues regarding language and ethnicity in the context of the integration of plurilingual repertoires in the education system. I end this chapter with a short section on the challenges and opportunities involved and some concluding remarks.

Language Diversity in Education in Colombia

Colombia is a Spanish-speaking country whose 1991 constitution (Constitución Política de Colombia, Art. 7) recognises it as a multi-ethnic and pluricultural nation, acknowledging the existence and officiality of 65 Indigenous languages, two creole languages, Colombian sign language, and the Rom language, which is a variety of the Romani language (Indo-European family) from North India spoken by nomadic people. Despite the constitutional recognition of these languages, Colombia's educational policy prioritises English within the official policy of bilingualism (de Mejía et al. 2023), instantiated through two programmes: *Colombia Bilingüe* and *Programa Nacional de Bilingüismo*. Over the years, English teaching and learning in the context of these programmes have failed to provide students with the 'professional and academic skills needed to succeed in the global market' (Ortega 2023, pp. 133–134), as the programmes essentially delinked linguistic content from the cultural and social realities of students around the country (González 2010; Usma Wilches 2015). Consequently, in the last 20 years, researchers in Colombia have advanced plural approaches that acknowledge and embrace linguistic and cultural diversity (Sánchez Moreano et al. 2023; de Mejía et al. 2023) in light of developments in the field of multilingualism in both the Global South and North (Léglise 2017; Martin-Jones and Martin 2017; García et al. 2016; May 2014).

Prior to the 1991 constitution, Columbia recognised only Spanish as the official language. Isolated initiatives for the inclusion of Indigenous languages in schools emerged from the 1930s onwards, supported by Indigenous movements, especially in the region of Cauca in Southwest Colombia (López and Wolfgang 1999). Later, in the 1960s and 1970s, alternative educational programmes emerged with proposals for new trends for bilingual Indigenous education and aimed for higher quality and more equitable education (López and Wolfgang 1999). Later, the focus shifted from only teaching and learning linguistic content to also considering the ancestral knowledge and cultures associated with Indigenous languages and the active participation of community members in the design of intercultural bilingual programmes. Thus, at the beginning of the 1970s, the Regional Indigenous Council of Cauca (Consejo

Regional Indígena del Cauca: CRIC) proposed a new model—*Educación propia* or *Indigenous People's Education*—which aimed to make room for local and cultural forms of knowledge that are not recognised in traditional homogenising schooling (González Terreros 2012). From this model emerged the Bilingual and Intercultural Education Programme (PEBI) under the fundamental principles of the CRIC: unity, land, culture, and autonomy (Universidad Autónoma Indígena Intercultural 2019).

However, instead of promoting bottom-up initiatives based on the philosophy of *Educación propia*, in the mid-1980s, official policies aimed at including diversity in education focused mainly on ethno-education: a government project and state policy (República de Colombia 1994, Art. 7) seeking to 'normalise and regularise most of the educational experiences initiated by Indigenous organisations in the previous decade under the ideological banners of ethno-development and interculturality' (Calvo Población and García Bravo 2013, p. 345). The Ministry of Education is responsible for establishing the framework for its implementation. However, even though ethno-education was developed to empower Indigenous communities, its prospects for success have been disappointing, as it became a form of hegemonic discourse linked to broader processes of homogenisation, capitalism, and globalisation (Cardona and Echeverri Restrepo 1997).

Despite the recognition of Colombia's cultural and linguistic diversity and the promotion of new trends in multilingualism and ethno-education, the paradox of a historically homogenising educational model (Guerrero Rivera and Soler Castillo 2020) continues to have important effects on minority groups. Further, the essentialisation of language, culture, identity, and native speakerism has influenced language practices and policies the world over (O'Rourke and Ramallo 2011; Rampton 1990; Pujolar and O'Rourke 2022; Muni Toke 2013) and the desire for educational autonomy from Indigenous organisations raise fundamental problems and tensions regarding ethno-education (Valencia Giraldo et al. 2023).

For example, ethno-educators facilitate the transmission of Indigenous peoples' traditional knowledge, and the content of ethno-educative practices is organised around activities such as carving, weaving, looking for natural dyes, and designating them in Indigenous languages, describing what nature provides, raising awareness among students about 'eco-tourism' practices, respect for the sacred territory, horticultural practices, and environmental care. According to the Law of Native Languages (República de Colombia 2010), this transmission should be mandatorily carried out in students' languages or at least be based on the use of students' languages. However, many ethno-educators do not speak the same language as their students and carry out their teaching in Spanish, especially in the areas of science, technology, engineering, and maths.

According to some ethno-educators, the difficulty of implementing ethno-education programmes in schools in terms of means, resources, and teacher training leads many teachers to abandon traditional Indigenous education and focus more on Western education (Sánchez Moreano 2019). Furthermore, the very physical layout of classrooms in many institutions follows a Western-style classroom logic, where the teacher occupies a predominant place at the front while students are homogeneously distributed at the back. This spatial configuration does not seem to favour the practice of cultural activities specific to ethnic groups, which usually take place in *malokas* (traditional Indigenous housing where the transmission of knowledge and culture takes place), *chacras* (the Quichuan word for a small parcel of land used for agriculture), rivers, and other cultural and spiritual spaces of socialisation. The primary consequence is that Indigenous languages are not included in practice in the school curriculum, and the forms of knowledge associated with the Indigenous languages of the students are no longer transmitted from one generation to another in this institutional setting. Furthermore, the general discourse among teachers and educators suggests that students do not show interest in talking in, preserving, or revitalising their languages at school. This is due to broader shared ideologies that suggest that Spanish and international languages can lead to social and economic success and progress (Mahecha Rubio 2011). Fortunately, Columbian laws 1381 (República de Colombia 2010) and 2246 (República de Colombia 2023) constitute a step in somewhat acknowledging the role of ethno-educators and Indigenous leaders in the construction of Indigenous educational systems.

Context of the Study

This study is set in a region where processes of invisibilisation and the marginalisation of these Indigenous populations and their specificities have taken place sociohistorically in favour of extractivist interests (da Silva Guevara 2023). As a consequence, forms of ancestral Indigenous knowledge and multiple cosmovisions (e.g., the different ways in which individuals or social groups perceive the world) that can contribute to the sustainable use of resources and economic sustainability have been ignored in the construction of the Colombian nation-state (da Silva Guevara 2023).

As in any other space of sociocultural diversity characterised by language ideologies (Irvine and Gal 2001), the linguistic resources (e.g., languages, dialects, varieties, repertoires, and registers) of the speakers who inhabit the region do not circulate equally in the various discursive spaces and social networks. This translates into two major problems. On one hand, access to health, education, social services, the labour market, the local and global economy, and other crucial aspects of citizenship (e.g., sustainable development and its benefits for society and the environment) is unequal. On the other hand, the intergenerational transmission of ancestral/biocultural knowledge (Maffi 2018) that characterises the relationship between ethnic groups and nature is interrupted by the ideological weight of majority and prestige languages, for example, Spanish, English, and Brazilian Portuguese. Further, due to military colonisation and actions taken by missionaries, Spanish language teaching played an important role in the assimilation of Indigenous populations to the Colombian nation (da Silva Guevara 2023) and was positioned as the 'language for all at the expense of the country's inherent multilingualism' (Valencia Giraldo et al. 2023, p. 16).

Prevalent language ideologies systematically exclude from educational systems (and from all other aspects of citizenship) minority and Indigenous languages, including the knowledge of the natural world that these languages convey. Some of the concrete consequences of these ideologies are that while Spanish is the language of instruction, schools are primarily multiethnic; that English is the international language that students wish to learn first; that only prestige languages, such as Spanish and Brazilian

Portuguese, and to a lesser extent the mixture of both (i.e. *Portuñol*), are vindicated by non-Indigenous populations (Chinellato Díaz 2021); and that all state services are offered exclusively in Spanish, for example, hospital care requires patients to have proficiency in Spanish.

In the school setting, these ideologies are visible in the attitudes and representations of students towards their ethnicities and Indigenous languages. This is evident when, for instance, Daniel (a student) does not accept his ethnicity and says, 'I don't like to accept that I'm Tikuna', or when Ronaldo prefers international languages, such as English, 'English sounds cooler than Tikuna', and shows some indifference towards the loss of his ethnic language, adding, 'I don't care if the language disappears'. These negative representations and attitudes towards Indigenous ethnicity and languages are common in this context. Sociolinguists are cognisant that such attitudes are the consequence of dominant monolingual ideologies that suggest that international, majority, and prestige languages are more important than Indigenous languages. As a result, the use of the Indigenous languages in the Amazon is alarmingly low (Acosta Muñoz et al. 2020), and the integration of plurilingual repertoires, as part of the United Nations Sustainable Development Goals for inclusive educational systems and sustainability, constitutes a major challenge for researchers and educators (Léglise 2017) in this context of ideologised language and teaching practices.

Ethnography in Plurilingual Settings

This study is based on observations of ethno-educative teaching practices, informal field interviews, awareness-raising workshops, and fieldnotes collected during several periods of ethnographic fieldwork from 2018 to 2024 (excluding the pandemic years 2020–2021) in a multiethnic school in the periphery of Leticia. Ethnographic observation of teaching practices in multilingual institutional settings is one of the most common methods used in language and education studies (Heller and Martin-Jones 2001; Heller 2006; Martín Rojo 2010; Patiño-Santos 2018; Pérez-Milans and Patiño-Santos 2014). One of the techniques used was participant observation, which consisted of taking part in the teaching activities proposed by teachers while observing the development of the activities and the interactions between teachers and students.

The observations were mainly conducted in classes on culture and eco-tourism because two of the teacher collaborators oversaw these two themes. We had the opportunity to indirectly observe the informatics and mathematics courses without participating in the activities. In the culture classes, for example, we participated in activities such as the use of family vocatives in different languages, which the students had to investigate with their parents and grandparents in their communities, or narratives pertaining to the preparation of local meals in different Amazon communities. Outside these classroom activities, we conducted informal field interviews with various school actors, such as students, teachers, and the principal. Although these interviews were not structured, they allowed us to address specific issues, such as the structural problems of the school and individual cases of student problems (e.g., drug use, family problems, intention to drop out).

We also conducted a series of workshops with five focal groups of students to raise awareness of environmental issues, such as food supply systems, deforestation (DEF), and eco-tourism (ECOT). These workshops were carried out under the Multilingualism, Sustainability, and Education in Amazonia research programme. Two research assistants animated most of the workshop sessions. Fifty-two students participated in the workshops, all of whom attended secondary school, were aged 15 to 19, and belonged to seven ethnic groups: Tikuna, Yukuna, Uitoto, Bora, Yagua, Cocama, and Ocaina.

We also conducted language portrait workshops, which are an analytical tool used in linguistic reflexivity activities in educational contexts to raise awareness of language values and multilingualism through the representation of embodied multimodal language repertoires (Busch 2015). This tool is used extensively by the research group Heteroglossia in the fields of sociolinguistics, applied linguistics, and discourse analysis. Research by Heteroglossia draws on biographical and ethnographic approaches to language use in heteroglossic contexts. Language portraits consist of a visual representation of the biographical experience of language and a narrative or interaction based on this visual representation. In language portraits, the image and language are reciprocally associated and interdependent, creating a socially situated and interactionally achieved unit of meaning. The image allows the initiation of conversation, working as a point of reference in the conversation and enabling participants to reflexively account for their language practices, representations, and ideologies (Busch 2015).

All sessions were recorded, transcribed, and completed with field notes. The transcriptions were qualitatively analysed following a conversation and interactional framework (Mondada 2002) of language practices and discourses. Various forms of data (participant observations, informal interviews, workshops) were used in this study, and all personal data were anonymised (e.g., pseudonyms were used). Owing to this combination of methods, we identified the limitations and potentialities of introducing Indigenous languages into the school curriculum. Our reflections in the section 'Issues and challenges in the integration of plurilingual repertoires in a multi-ethnic schools' show the limitations, potentiality, and temporalities whereby plurilingual-based language practices in school can be shaped within the context of epistemic and environmental justice.

Issues and Challenges in the Integration of Plurilingual Repertoires in a Multi-Ethnic School

Acknowledging and Assigning Student Ethnicity

One of the main issues affecting the integration of plurilingual repertoires in the multi-ethnic school in terms of addressing environmental and sustainability issues is acknowledging and assigning students' ethnicity in pedagogic activities. We observed this during most of the activities of the class on culture in the multi-ethnic school in Leticia. In terms of the students acknowledging their identities, we observed that most of them did not feel comfortable when the ethnicity question was at stake in any kind of activity in which their participation was needed, and they tended to reject their own ethnicity, as shown in Example 1.

(1) Interactions in a multi-ethnic classroom in a school in Leticia (Wang and Sánchez Moreano, in press)

23. **Santiago** (...) ¿Y por qué no pusiste (el tikuna)? Ah, pero hablas tikuna mejor dicho
[And why didn't you include (the Tikuna language)? Ok, but you speak Tikuna, right?]
24. **Daniel** Eeh, no, no no hablo ni entiendo
[Euh, no, no, I don't speak, nor do I understand]
25. **Santiago** no hablas ni entiendes tikuna
[You don't speak, nor do you understand Tikuna]
26. **Daniel** No, por eso no la puse
[No, this is why I didn't include it]
27. **Santiago** Okey, ¿pero pero hace parte de tu vida?
[Ok, but, but it forms a part of your life?]
28. **Daniel** Aja, hace parte de mi vida
[Aha, it forms a part of my life]
29. **Santiago** Ah bueno, pero la consigna era precisamente poner las que hacen parte de la vida de uno ¿no?
[Aah ok, but the instruction was to include those which form a part of your life, right?]
30. **Daniel** (Laughing in a shy manner).
31. **Santiago** Aahh

32.	**Daniel**	No profe en ese caso no me gusta aceptar que soy tikuna profe, no me gusta [No teacher, the thing is that I don't like to accept that I am Tikuna teacher, I don't like it]

In the above interaction, the student explains the result of a task regarding the languages that make up part of his life. In line 23, the researcher asks him why he did not include the Tikuna language as part of the language repertoire of his life. In lines 24 and 26, he says that he does not understand the language in question and that for this reason, he did not include it in the task. Later, in 32, the student reinforces his stance, saying that he does not like to accept that he belongs to the Tikuna community.

This example also shows the ethnolinguistic assignation by the researcher. In sociology, the attribution of a specific identity to individuals or social groups is called *assignation identitaire* (Bessone et al. 2015; Duvoux 2017; Landais 2011); in sociolinguistics and the sociology of language, it is called *assignation linguistique* (Adam Maillet 2021; Billiez 2002; Ellouzi 2022). In the Anglo-Saxon literature, we speak of racialisation (Rosa and Flores 2017; Flores and Rosa 2023) and ethnolinguistic cornering (Møller 2021, 2022, 2023). These terms provide a toolkit for the analysis of the negative consequences (e.g., symbolic violence or domination; Bourdieu 2001) produced by the act of assigning a linguistic identity to people, often from minority groups that do not speak the language of the majority group, thereby denying their specificities and sociocultural backgrounds (Adam Maillet 2021).

The concept of ethnolinguistic cornering is based on Silverstein's (1979) work on linguistic ideologies, which describe interactional sequences in which social actors explicitly reproduce ethnolinguistic assumptions, that is, the close connection between a language and an individual or group of people, often with a negative charge (Blommaert et al. 2012). Observed in social interaction, ethnolinguistic cornering may create boundaries and attribute otherness, for example, a specific identity based on ideological assumptions (Møller 2021). Although the concept has not been extensively used in the literature, it is a suitable analytical tool for systematic analyses of social interactions in the classroom.

In Example 1, the researcher performs ethnolinguistic cornering in line 23 and reinforces it in lines 25 and 29, creating an uncomfortable situation for the student, reflected in his shy smiles (line 30) and the need for him to justify his non-affiliation with the Tikunas. However, ethnolinguistic assignation was also observed in many other classroom activities, often on the part of educators. In Example 2, during the culture class, the teacher refers to the students based on perceptions of their ethnicity, probably according to the declared ethnicity found in the student registration list and knowledge of the group from the beginning of the school year.

(2) Classroom observations and fieldnotes

> During the sequence, the teacher proposed a collective activity to document family vocatives (mom, dad, uncle, aunt, etc.), as expressed in the students' language. The students had to investigate the equivalent of these vocatives with members of their community and family. The activity raised considerable interest among the students, and all seemed to have undertaken the task within their communities. During the socialisation activity, the teacher created several columns on the blackboard, each one referring to some of the more populated ethnic groups and their languages in the zone (e.g. Cocama, Tikuna, Uitoto, Bora, Yagua, etc.). To promote students' participation, the teacher started by saying, 'levanten la mano los cocama' [raise your hands] in Cocama. Following five seconds of silence, only two students raised their hands shyly. She invited them to write down on the blackboard the Cocama equivalents for father and mother. The teacher then continued, '¿donde están los tikunas, los tikunas por allá, en la segunda fila?' [where are the tikunas, the tikunas over there, in the second row?] – pointing at four students sitting in the second row.

The teacher continued the activity by calling each of the ethnic groups in turn. Throughout the activity, the teacher took for granted the students' ethnicity, at times eliciting discomfort, vulnerability, and even sociolinguistic discrimination. The teacher's attitude showcased an unconscious process of ethnolinguistic cornering that was sometimes reproduced by some of the students. When the teacher said, 'anyone else' (addressing the Yukuna students), one of the students said, 'Jefferson!' pointing towards one student who had not raised his hand or gone to the blackboard because he did not feel comfortable with the activity and probably with the identity assignation. The teacher continued the activity, sometimes forcing the ethnolinguistic assignation 'quién mas (who else, pointing towards another student)?' The teacher asked one girl, 'you are Tikuna (Uitoto? Yukuna?)?' The student nodded with some acknowledgement, with a certain sense of discomfort about having to admit belongingness to the Yukuna group. The teacher then concluded the sequence, adding, 'no sientan verguenza de su etnia (don't be ashamed of your ethnicity)!' while another student proceeded 'voluntarily' to the blackboard, 'qué orgullo (I'm proud of you!)!' said the teacher, trying to encourage the student.

Based on informal conversations with students after the class, they were not generally appreciative of such activities. The two examples discussed above show that even if the ethno-educative teaching practices were intent on including students' language and culture in the classroom, the result may have been problematic, with negative effects on student learning. For example, during socialisation activities, students may experience discomfort, vulnerability, and sociolinguistic discrimination, which can unintentionally contribute to the performance of symbolic violence (Bourdieu 2001). The activities proposed may not achieve the expected outcome, for example, including students' linguistic repertoires in the classroom. Further, these practices can reinforce students' apprehension about being exposed to discrimination, resulting in loss of interest and potential abandonment of their languages and cultures, at least in the institutional context.

Raising Awareness of the Value of Students' Linguistic Resources: Towards Linguistic Citizenship in School

In the section 'Issues and challenges in the integration of plurilingual repertoires in a multi-ethnic school', we saw that students distanced themselves from their ethnicities (Example 1) and that the ethnolinguistic cornering performed by teachers, educators, and even researchers vis-à-vis students (Example 2) reinforced the negative representations and attitudes that students have of their ethnic languages. Both situations may end up in loss of interest in the preservation/transmission of Indigenous language practice. These negative outcomes, which impact the complex exercise of introducing students' languages and repertoires in school teaching practice, led us to propose a series of participative workshops to raise awareness of multilingualism. The workshops have been taking place since 2018 in the school, especially in the secondary school (Fig. 10.1, on the left), and students have produced numerous language portraits (Fig. 10.1, on the right) that have been analysed by drawing on earlier research by Sánchez Moreano (2019) in terms of representations and the lived experience of language (Busch 2015).

Working with language portraits in the school enabled a great number of students to change their views regarding the languages they speak at home, despite not being visible at school. Example 3 shows that in the case of Sheila, the Tikuna language has its place at home and that even

Fig. 10.1 On the left: Language portrait workshop (Leticia in 2019). On the right: A data set of language portraits (August 2022). Photos taken by the author

if she understands it only marginally, she prefers Tikuna to Spanish. Example 4 shows that not only does the student prefer the Tikuna language but that she also has a desire for openness to transmit her family language to all those who would like to learn it.

(3) Language portrait workshops, August 2022

> Sheila: 'El tikuna, lo hablo pero poquito, entiendo un poquito. Mi abuela me habla en tikuna. Me gusta más la lengua tikuna que el español'.
> [The Tikuna language, I speak it a little bit, I understand it a little bit. My grandma talks to me in Tikuna. I prefer Tikuna to Spanish.]

(4) Language portraits workshops, August 2023

> Shirley: 'Mi mamá me habla en tikuna, me gusta hablar para más aprender y enseñar a otros'
> [My mother talks to me in Tikuna; I like speaking to learn more and teach others].

Language portraits have the potential to transform participants' realities, developing positive values of the language/s in their linguistic repertoires. We consider that raising awareness of the value of the languages in students' lives in the multi-ethnic school in Amazonia was a mandatory first step before trying to introduce Indigenous languages into the school curriculum and teaching practices. Against the generalised teachers' discourse that students do not want to learn their languages, we found it important to tackle this 'lack of interest' within the school community to raise awareness of the values of students' plurilingual repertoires. The idea was to say that it is not only a matter of suggesting to students that they should have an interest in their languages and cultures. We cannot request that they speak their languages. Instead, it is a matter of raising awareness about the fact that their linguistic repertoires are plural and diverse and that they, as social actors, have the capacity to use them to create meaning and express their voices. In other words, the idea is to say that their multi-ethnic voices have value in the same way that as so-called 'prestige' languages.

This echoes the concept of linguistic citizenship as an engagement with diversity and marginalisation through a sort of 'politics of the ordinary' (Williams and Stroud 2015), which determines language practices in everyday life, outside of institutional settings. However, the school examined in this study is an institutional setting constrained by ideologies of normativity and the potential production of social inequalities. For example, in addition to the language ideologies that promote Spanish and English as prestige languages, the current school curriculum considers that the language and culture that should be predominantly promoted are those of Tikuna because the majority of students are Tikunas. However, focusing on the Tikuna language and culture systematically excludes all other minority ethnic groups that attend the school. Therefore, despite the good intentions around the focus on the Tikuna language and culture, this imposition from above echoes what Williams and Stroud (2015) called passive forms of linguistic citizenships: voices of the marginalised remain silent to centralised forms of decision-making. Consequently, despite the restrictive and constraining nature of the institutional setting, the students attending this school claim opportunities to express themselves and exist as full citizens in a plurilingual, pluri-diverse setting, as shown in Examples 5, 3, and 4.

(5) Eco-tourism workshops, April 2024

> Pues yo, pues en mi Pueblo, tierra conozco la danza, y pues como dijo mi compañero, la lengua. Y, pues a mí me gustaría aprender eso. Porque es muy bonito y mi mamá lo hace, mi mamá lo habla. Pues yo obviamente, yo no la entiendo. Y, pues ella me trata de enseñar, pues sí, ahí voy aprendiendo paso a paso y es muy bonito hablar la verdad.
> [Well, in my village, in my territory, I know the dance, and as my colleague said, the language. And, well, I would like to learn that (the language). Because it is very beautiful. My mother does it, my mother speaks it. Well, obviously, I don't understand her. And, well, she tries to teach me, so yes, I'm learning step by step, and it's very nice to speak it, really.] (Translated by the author)

We consider that raising awareness is not a sufficient strategy to include languages and plurilingual repertoires in school teaching practices. Linguistic citizenship, as a praxis, can be applied to multilingual contexts, such as the Amazon Trapezium, where languages of unequal status co-exist, with the aims of eliminating inequalities between groups, promoting social justice (Williams et al. 2022), and contributing to addressing environmental challenges, such as nature preservation and the construction of sustainable futures.

In addition to these reflections, any research about multilingualism, education, and sustainability should rely on four principles that underpin the relationality between language, people, and nature: (1) ancestral knowledge about nature encoded in Indigenous languages is crucial to understanding biodiversity and its sustainable management; (2) Indigenous voices have important insights to contribute to the epistemic dialogue on global environmental protection and sustainability (Harrison 2019; Wehi et al. 2024); (3) plurilingual practices, as a human and praxeological activity (Boutet 2016), are intrinsically linked to how people and nature relate to each other; and (4) pluralistic approaches to sustainability and biodiversity sciences should include multilingualism in sources and processes, as language plays a central role in people's lives and their relation to society and nature (Droz et al. 2023).

Discussion

Our chapter relied on the first stage of a broader agenda towards integrating Indigenous languages and the knowledge associated with them in the epistemic dialogue around nature protection through education. In the vein of social and environmental justice approaches from the ground up (Boaventura de Sousa 2012; Makoni et al. 2023), our agenda addresses not only the integration of Indigenous voices into education and sustainability but also understanding of the consequences of language and sustainability ideologies of power and control (Milroy 2001; Blackledge 2005; Holz 2018). In fact, these ideologies, modern forms of the coloniality of nature (Lander 1999), and Euro-American ideas of 'wild' nature (Pascual et al. 2021) have had devastating effects on Amazonian Indigenous communities, languages, and cultures.

Indeed, the integration of multilingual practices into education systems remains a challenge for research in the Global South, as education systems are caught between globalisation and colonisation (Léglise 2017). To address this challenge and create impact, Multilingual Education for Social Justice (Mohanty et al. 2009) suggests that the use of multiple languages in education can lead to social and epistemic justice, as all languages and language varieties have a legitimate place in the world. Furthermore, by mobilising decolonial approaches to sustainability and nature (Arteta Melgarejo et al. 2021), we expect to go beyond the economic aspects of sustainable development towards pluralistic approaches to biodiversity and its conservation based on the socio-natural networks built and defended by ordinary people (Escobar 2014).

By focusing on a multi-ethnic school, we chose an institutionalised setting immersed within the complexities of a territory, Amazonia, which is subject to modern forms of coloniality (Mignolo and Walsh 2018; Descola 2005). In this context, developing and applying the concepts of linguistic citizenship and environmental justice are central to the empowerment of new generations of social actors in preventing the reproduction of colonial futures (Stein et al. 2023). We maintain that one way of empowering young people from the school is to explore new paths of collaboration between multilingualism as a field of study and environmental/sustainability sciences. While multilingualism and its implications for society and education have been widely explored in the humanities and social sciences in both the Global North and South (May 2014; García et al. 2016; Martin-Jones and Martin 2017), language and multilingualism remain an opaque concept in socio-ecological and environmental research, especially in complex contexts, such as the Amazonia region where the forest, fauna, and multi-ethnic Indigenous groups, along with their languages, are in grave danger of extinction (Grinevald 1998).

Opening these paths of collaboration may lay the groundwork for a new understanding of the relationship between people and nature, the empowerment of young people from communities, and the development of multi-actor participatory methodologies. We believe that our reflections can provide guidance to non-Indigenous local and global communities, policymakers, SMEs interested in the bioeconomy, teachers, private/public funders, and stakeholders engaged in sustainability in the rest of the Amazonia region on multilingualism in education for sustainability. Finally, we maintain that our reflections, which are aligned with the United Nations Sustainable Development Goals, can contribute to the broader field of education for sustainability, which focuses on inter- and transdisciplinary modes of producing and circulating knowledge and an epistemic dialogue for integrating diverse ways of knowing.

Acknowledgements We would like to thank our social geographer research assistants Lina Vanesa Sanchez and Juan Sebastian Valencia for their valuable work in the Multilingualism, Sustainability, and Education in Amazonia project (MSEA 2023–2024), funded by The Open Societal Challenges Programme at the Open University.

References

Acevedo, L.M.G. 2011. ¿Cultura para consumir? Los yagua y el turismo cultural en el Trapecio Amazónico. *Revista Colombiana de Antropología* 47, no. 1 (August 20): 113–136.

Acosta Muñoz, L.E., D. Mendoza Hernández, P.E. De La Cruz Nassar, and U.G. Murcia García. 2020. *Indicadores de Bienestar Humano Indígena (IBHI): Primer reporte sobre el estado de los modos de vida y territorios de los pueblos Indígenas del departamento de Amazonas*. Colombia: Instituto Amazónico de Investigaciones Científicas SINCHI.

Adam Maillet, M. 2021. "Parlez votre langue!". Injonctions plurilingues et violences symboliques faites aux élèves allophones. Recherches en didactique des langues et des cultures. *Les cahiers de l'Acedle* 18–2, no. 18–2 (September 28). Available from: https://journals.openedition.org/rdlc/9175. Accessed October 2, 2024.

Arteta Melgarejo, X.C., A.C. Molina Padilla, P.L. Oliveros Eusse, and L. Vasquez-Peñaloza. 2021. Colonialidad de la naturaleza: Aspectos decoloniales para el debate sobre el desarrollo sostenible. *Encuentrosno.* 14 (August 25): 288–300.

Bessone, M., A. Cukier, C. Lazzeri, and M.-C. Willems. 2015. Identités et catégorisations sociales. *Terrains/Théoriesno.* 3 (October 26). Available from: https://journals.openedition.org/teth/503. Accessed October 2, 2024.

Billiez, J. 2002. De l'assignation à la langue d'origine à l'éveil aux langues: Vingt ans d'un parcours sociodidactique. *Diversité* 130 (1): 87–101.

Blackledge, A. 2005. *Discourse and power in a multilingual world*. Amsterdam; Philadelphia, PA: John Benjamins Pub. Available from: http://public.eblib.com/choice/publicfullrecord.aspx?p=622468. Accessed October 2, 2024.

Blommaert, J., S. Leppänen, and M. Spotti. 2012. Endangering multilingualism. In *Dangerous multilingualism*, ed. J. Blommaert, S. Leppänen, P. Pahta, and T. Räisänen, 1–21. London: Palgrave Macmillan UK. Available from: http://link.springer.com/10.1057/9781137283566_1. Accessed October 2, 2024.

Boaventura de Sousa, S. 2012. Public sphere and epistemologies of the South. *Africa Development* 37 (1): 43–67.

Bourdieu, P. 2001. *Langage et pouvoir symbolique*. Paris: Seuil. Available from: https://www.editionspoints.com/ouvrage/langage-et-pouvoir-symbolique-pierre-bourdieu/9782757842034. Accessed October 2, 2024.

Boutet, J. 2016. *Le pouvoir des mots*. Nouvelle Édition. Paris: La Dispute.

Brondízio, E.S., Y. Aumeeruddy-Thomas, P. Bates, J. Carino, Á. Fernández-Llamazares, M.F. Ferrari, K. Galvin, V. Reyes-García, P. McElwee, Z. Molnár, A. Samakov, and U.B. Shrestha. 2021. Locally based, regionally manifested, and globally relevant: Indigenous and local knowledge, values, and practices for nature. *Annual Review of Environment and Resources* 46, no. 1 (October 18): 481–509.

Busch, B. 2015. Expanding the notion of the linguistic repertoire: On the concept of *spracherleben*—The lived experience of language. *Applied Linguistics* 38, no. 3 (July 24): 340–358. https://doi.org/10.1093/applin/amv030.

Calvo Población, G.F., and W. García Bravo. 2013. Revisión crítica de la etnoeducación en Colombia. [Critical review of ethnic education in Colombia]. *Historia de la educació* 32: 343–360. Available from: http://hdl.handle.net/10366/130918. Accessed November 12, 2024.

Cardona, M., and J.A. Echeverri Restrepo. 1997. Etnoeducación y cultura: Elementos para una caracterización de la educación Indígena en el departamento del Amazonas. *Boletín De Antropología* 11 (28): 71–92.

Chinellato Díaz, A. 2021. El portuñol en la triple frontera amazónica: Del déficit al translenguar. *LaborHistórico* 7, no. 1 (April 30): 45–69.

Constitución Política de Colombia. 1991. Art. 7.

Descola, P. 2005. *Par-delà nature et culture*. Paris: NRF: Gallimard.

Droz, L., M. Brugnach, and U. Pascual. 2023. Multilingualism for pluralising knowledge and decision making about people and nature relationships. *People and Nature* 5, no. 3 (March 28): 874–884. https://doi.org/10.1002/pan3.10468.

Duvoux, N. 2017. Chapitre III—Les formes de l'assignation identitaire. *Que sais-je?* (September 11): 56–80.

Echeverri Restrepo, J.A. 2023. Leticia indígena: Construcción territorial Indígena en la ciudad. *Mundo Amazónico* 14, no. 1 (January 3): 49–79.

Ellouzi, H. 2022. Le processus d'assignation identitaire à l'œuvre dans l'enseignement d'une langue « pas vraiment étrangère » en question. Thesis, Paris: Paris 3. Available from: https://theses.fr/2022PA030020. Accessed November 12, 2024.

Escobar, A. 2014. *Sentipensar con la tierra: Nuevas lecturas sobre desarrollo, territorio y diferencia*. Primera edición. T Colección Pensamiento Vivo. Medellín, Colombia: Ediciones Unaula.

Flores, N., and J. Rosa. 2023. Undoing raciolinguistics. *Journal of Sociolinguistics* 27, no. 5 (November): 421–427. https://doi.org/10.1111/josl.12643.

García, O., N. Flores, and M. Spotti, eds. 2016. *Bilingualism, multilingualism, globalization, and superdiversity*. Oxford: Oxford University Press.

Gasché, J., and N. Vela Mendoza. 2011. *Sociedad bosquesina. Vol. I.* II vols. Iquitos, Perú: Instituto de Investigaciones de la Amazonía Peruana (IIAP).

González, A. 2010. English and English teaching in Colombia: Tensions and possibilities in the expanding circle. In *The Routledge handbook of world Englishes*, ed. A. Kirkpatrick, 354–374. London: Routledge.

González Terreros, M.I. 2012. La educación propia: Entre legados Católicos y reivindicaciones étnicas. *Pedagogía y Saberes* no. 36 (April 1) https://doi.org/10.17227/01212494.36pys33.43.

Grinevald, C. 1998. Language endangerment in South America: A programmatic approach. In *Endangered languages*, ed. A.L. Grenoble, and J.W. Lindsay, 124–160. Cambridge: Cambridge University Press. https://doi.org/10.1017/CBO9781139166959.007.

Guerrero Rivera, J., and S. Soler Castillo. 2020. Representación de los Indígenas en las leyes generales de educación de Colombia. *Foliosno.* 52 (December): 71–86.

Harrison, K. 2019. Languages, plants, and people. *Language Magazine* (March 14): 28–32.

Heller, M. 2006. *Linguistic minorities and modernity: A sociolinguistic ethnography*. London; New York: Continuum.

Heller, M., and M. Martin-Jones (eds.). 2001. *Voices of authority: Education and linguistic difference. Contemporary studies in linguistics and education.* Vol. 1. Connecticut, Ablex.

Holz, P. 2018. *Towards a new social order? Real democracy, sustainability & peace. Vernon Series in Sociology.* Wilmington, Delaware: Vernon Press.

Irvine, J.T., and S. Gal. 2001. Language ideology and linguistic differentiation. In *Regimes of language. ideology, politics and identity*, ed. P.V. Kroskrity, 35–83. Santa Fe, Oxford: School of American Research Press, James Currey.

Landais, P. 2011. Du « social » à l'« identitaire » ? La question de l'assignation identitaire des immigrants et de leurs descendants. In *L'Identité, entre ineffable et effroyable*, 50–61. Paris: Armand Colin.

Lander, E. 1999. ¿Conocimiento para qué? ¿Conocimiento para quién? Reflexiones sobre la universidad y la geopolítica de los saberes hegemónicos. *Estudios Latinoamericanos* 7 (12–13): 25.

Léglise, I. 2017. Multilingualism and heterogeneous language practices: New research areas and issues in the Global South. Trans. by Linda Gardiner. *Langage et société* 160161, no. 2 (May 24): 251–266.

López, L.E., and K. Wolfgang. 1999. La educación intercultural bilingüe en américa latina: Balance y perspectivas. *Revista Iberoamericana de Educaciónno 20. OEI: 50 Años de Cooperación / OEI: 50 Anos de Cooperação*: 17–84.

Maffi, L. 2018. Biocultural diversity. In *The International Encyclopedia of anthropology*, ed. H. Callan, 1–14. 1st ed. Wiley. Available from: https://onlinelibrary.wiley.com/doi/10.1002/9781118924396.wbiea1797. Accessed October 2, 2024.

Mahecha Rubio, D. 2011. Escuela y multiligüismo en Amazonia: Un desafío contemporáneo. In *Amazonia Colombiana: Imaginarios y realidades. Memorias de la Cátedra Jorge Eliecer Gaitán,* ed. J.A. Echeverri Restrepo, and C. Pérez Niño, 293–309. Bogotá: Universidad Nacional de Colombia.

Makoni, S., A. Kaiper-Marquez, M. Madany-Saá, and B. Edem Antia, eds. 2023. *Foundational concepts of decolonial and southern epistemologies. Global Forum on Southern Epistemologies 3.* Bristol Jackson: Multilingual Matters.

Martín Rojo, L. 2010. Chapter 3, Critical sociolinguistic ethnography in schools. In *Constructing inequality in multilingual classrooms*, 51–92. De Gruyter Mouton. https://doi.org/10.1515/9783110226645.bm.

Martin-Jones, M., and D. Martin, eds. 2017. *Researching multilingualism: Critical and ethnographic perspectives*. London, New York: Routledge.

May, S. 2014. *The multilingual turn. Implications for SLA, TESOL, and bilingual education*. New York & London: Routledge Taylor & Francis.

de Mejía, A.-M., S. Valencia Giraldo, and N. Miranda. 2023. Critical perspectives on language education in multilingual Colombia: An introduction. In *Language education in multilingual Colombia: Critical perspectives and voices from the field*, ed. N. Miranda, A.-M. de Mejía, and S. Valencia Giraldo, 1–12. New York: Routledge.

Mignolo, W.D., and C.E. Walsh. 2018. *On decoloniality: Concepts, analytics, praxis*. Duke University Press.

Milroy, J. 2001. Language ideologies and the consequences of standardization. *Journal of Sociolinguistics* 5, no. 4 (November): 530–555. https://doi.org/10.1111/1467-9481.00163.

Mohanty, Ajit K., M. Panda, R. Phillipson, and T. Skutnabb-Kangas (eds.). 2009. *Multilingual education for social justice: Globalising the local*. New Delhi: Orient Blackswan.

Møller, J.S. 2023. *Cornering—Om Uønskede Sprogidentiteter*. Internet publication. Available from: https://dialekt.ku.dk/maanedens_emne/cornering/. Accessed October 2, 2024.

Møller, J.S. 2021. Etnolingvistisk cornering. In *Sprogs Status i Rigsfællesskabet 2031*, ed. T. Kristiansen, and A. Holmen, C 13: 115–130. Københavnerstudier i Tosprogethed. Available from: https://cip.ku.dk/udvikling-og-forskning/forskning/studier_i_parallelsproglighed/Sprogs_status_i_Rigsf_llesskabet_2031_SAMLET.pdf. Accessed October 2, 2024.

Møller, J.S. 2022. *Ethnolinguistic cornering*. Seminar presentation presented at the Séminaire doctoral—Pratiques langagières—terrains, méthodes, théories, February 11, Paris.

Mondada, L. 2002. Pour une approche interactionnelle de la catégorisation des ressources linguistiques par les locuteurs. In *France, pays de contacts de langues*, ed. V. Castellotti, and D. de Robillard, 1: 23–35. Louvain-la-neuve: Cahiers de l'institut de linguistique de Louvain.

Muni Toke, V. 2013. Native speaker: From idealization to politicization. *Histoire Épistémologie Langage* 35 (2): 69–93. https://doi.org/10.3406/hel.2013.3458.

Ochoa Zuluaga, G.I. 2019. Coordinación diversificada y diálogo de saberes para cadenas de valor de turismo. *Turismo y Sociedad* 25 (June 10): 151–168. https://doi.org/10.18601/01207555.n25.08

O'Rourke, B., and F.F. Ramallo. 2011. The native-non-native dichotomy in minority language contexts: Comparisons between Irish and Galician. *Language Problems & Language Planning* 35 (2): 139–159. https://doi.org/10.1075/lplp.35.2.03oro.

Ortega, Y. 2023. Late capitalism and the commodification of English in Colombia: The shaping of language education policy and practice. In *Language education in multilingual Colombia. Critical perspectives and voices from the field*, ed. N. Miranda, A.-M. de Mejía, and S.V. Giraldo, 131–148. New York: Routledge.

Pascual, U., W.M. Adams, S. Díaz, S. Lele, G.M. Mace, and E. Turnhout. 2021. Biodiversity and the challenge of pluralism. *Nature Sustainability* 4, no. 7 (March 25): 567–572. https://doi.org/10.1038/s41893-021-00694-7.

Patiño-Santos, A. 2018. Introduction. *International Journal of the Sociology of Language 2018*, no. 250 (March 1): 1–10. https://doi.org/10.1515/ijsl-2017-0052.

Pérez-Milans, M., and A. Patiño-Santos. 2014. Language education and institutional change in a Madrid multilingual school. *International Journal of Multilingualism* 11, no. 4 (October 2): 449–470. https://doi.org/10.1080/14790718.2014.944532.

Pujolar, J., and B. O'Rourke. 2022. Theorizing the speaker and speakerness in applied linguistics and sociolinguistics. *Journal of Applied Linguistics and Professional Practice* 16, no. 2 (April 7): 207–231. https://doi.org/10.1558/jalpp.22760.

Rampton, B. 1990. Displacing the 'native speaker'. Expertise, affiliation and inheritance. *ELT Journal* 44, no. 2: 97–101. https://doi.org/10.1093/eltj/44.2.97.

de Rementería, I. 2011. Utilidad económica y ecológica de los cultivos ilícitos en el oriente de Colombia. In *Amazonia Colombiana: Imaginarios y realidades. Memorias de la Cátedra Jorge Eliecer Gaitán*, ed. J.A. Echeverri Restrepo, and C. Pérez Niño, 431–460. Bogotá: Universidad Nacional de Colombia.

República de Colombia. 1994. Ley 115. Ley general de educación.

República de Colombia. 2010. Ley 1381. Ley de lenguas nativas.

República de Colombia. 2023. Decreto 2246. Sobre la remuneración de los dinamizadores pedagógicos o educadores Indígenas.

Rosa, J., and N. Flores. 2017. Unsettling race and language: Toward a raciolinguistic perspective. *Language in Society* 46, no. 5 (November): 621–647. https://doi.org/10.1017/S0047404517000562

Sánchez Moreano, S. 2019. Repertorios lingüísticos y representación de la diversidad lingüística en Leticia (Colombia) y su periferia. *Visitas al Pationo.* 13 (July 21): 12–36. https://doi.org/10.32997/2027-0585-vol.0-num.13-2019-2313.

Sánchez Moreano, S., N. Miranda, I. Tejada-Sánchez, J. Barrera, A. González, O.C. Hernández, Y. Ortega, and Y. Cortés. 2023. Por una praxis educativa bi/pluri/ multi/lingüe en Colombia. *Notas de Política En Educaciónno.* 8: en línea. Available from: https://educacion.uniandes.edu.co/sites/default/files/educacion/Archivos/Notas%20de%20pol%C3%ADtica/2023/NotaDePolitica-Praxisbi%3Apluri%3Amulti.pdf. Accessed November 13, 2024.

da Silva Guevara, G. 2023. Más allá de la soberanía. La Amazonía y la geopolítica de la invisibilización, énfasis en Colombia y Brasil. *Estudios Políticosno.* 67 (June 23): 28–54. https://doi.org/10.17533/udea.espo.n67a02.

Silverstein, M. 1979. Language structure and linguistic ideology. In *The elements: A parasession on linguistic units and levels*, ed. P.R. Clyne, W.F. Hanks, and C.L. Hofbauer, 193–247. Chicago: Chicago Linguistic Society.

Stein, S., V. Andreotti, C. Ahenakew, R. Suša, W. Valley, N. Huni Kui, M. Tremembé, L. Taylor, D. Siwek, C. Cardoso, C.A. Duque, S. Oliveira Da Silva Huni Kui, B. Calhoun, S. Van Sluys, S. Amsler, D. D'Emilia, D. Pigeau, E. Bowness, and A. McIntyre. 2023. Beyond colonial futurities in climate education. *Teaching in Higher Education* 28, no. 5 (July 4): 987–1004. https://doi.org/10.1080/13562517.2023.2193667.

Universidad Autónoma Indígena Intercultural. 2019. *Retos y desafíos de la educación propia: Hacia los 50 años del CRIC presented at the 1era minga regional, nacional e internacional*, November 18, La Aldea, Popayán, Cauca.

Usma Wilches, J.A. 2015. From transnational language policy transfer to local appropriation: The case of the national bilingual program in Medellin, Colombia. Thesis. Ann Arbor, MI: ProQuest.

Valencia Giraldo, S., N. Miranda, and A.-M. de Mejía. 2023. Multilingual Colombia: Dimensions of linguistic and cultural diversity. In *Language education in multilingual Colombia critical perspectives and voices from the field*, ed. N. Miranda, A.-M. de Mejía, and S. Valencia Giraldo, 13–30. New York, NY: Routledge.

Wang, N., and S. Sánchez Moreano. In press. Language portraits in describing family language policy: How the activity setting shapes power dynamics.

Language and communication. Special issue: Ethnographic, biographic and genealogic approaches to complex linguistic situations: Methods in use with transnational families.

Wehi, P.M., H. Whaanga, T. Roa, and M.P. Cox. 2024. Woven languages: Understanding Indigenous socioecological systems. *Trends in Ecology & Evolution* 39, no. 10 (October 1): 881–884. https://doi.org/10.1016/j.tree.2024.08.004.

Williams, Q., and C. Stroud. 2015. Linguistic citizenship: Language and politics in postnational modernities. *Journal of Language and Politics* 14, no. 3 (August 21): 406–430. https://doi.org/10.1075/jlp.14.3.05wil.

Williams, Q., A. Deumert, and T.M. Milani, eds. 2022. *Struggles for multilingualism and linguistic citizenship. Multilingual Matters 173.* Bristol Jackson: Multilingual Matters.

11

Ecolinguistic Learning Modules for English Language and Linguistics

Salla-Riikka Kuusalu and Arran Stibbe

Introduction

As linguists and other language professionals, we are often faced with the realisation that the society we live in is an unjust and unsustainable one, but we may find it difficult to consider our own part in shaping that same society. Many of us are educators, who are considered modern change agents (Bürgener and Barth 2018; Weinberg et al. 2020), with the ability to affect the minds and actions of the student body that travels

Some sections of this chapter are adapted from Stibbe, Arran (2015) The search for new stories to live by: sustainability in the English Language and Linguistics curriculum. Advance HE. With permission from AdvanceHE.

S.-R. Kuusalu
School of Languages and Translation Studies, University of Turku, Turku, Finland

Department of Biology, University of Turku, Turku, Finland

A. Stibbe (✉)
University of Gloucestershire, Cheltenham, UK
e-mail: astibbe@glos.ac.uk

M. Maijala et al. (eds.), *Transformative Language Teaching for Sustainability*,
https://doi.org/10.1007/978-3-031-85493-4_11

through our lessons during our careers (e.g., Manuele 2015; Miller et al. 2018). Sometimes, these considerations give rise to a desire to do our part in changing, if not the world, then at least a considerable fraction of our immediate surroundings, in the hopes that this would lead to a cascade of positive changes in the future. Ecolinguistic Learning Modules, the topic of this chapter, have been created to answer the need to change education of English language and linguistics to embed critical thinking considering sustainability (Scott and Gough 2003; Öhman 2006; Lundegård and Wickman 2007, 2012; Osberg and Biesta 2010; Hasslöf and Malmberg 2015) and transformative action competencies (Frisk and Larson 2011; Odell et al. 2019) that support sustainable life.

Ecolinguistic Learning Modules are practical tools that help teachers guide the learner step by step on how to become more aware of the systems that affect our daily lives and the ways language use around us shapes how people think about the world, i.e., our collective, social cognition (see e.g., Augoustinos et al. 2014). Creating the modules required rethinking and reframing the concept of sustainability, especially related to sustainable development and the multiple uses and misuses of the term. Value systems shaping the thinking and actions of people are re-explained in the educational framework as 'the stories we live by' (Stibbe 2015), helping learners to grasp the complicated systems affecting the background. In addition, the learning modules are constructed to support the learners in developing responsive and responsible identities with practices of identity reflection (Anderson 2004).

This chapter introduces Ecolinguistic Learning Modules for English language and linguistics, first discussing alternative framings of sustainability, moving on to the theoretical background of the framework and its core concepts, and describing in more detail the structure of the modules and their applicability to other contexts. The chapter concludes by considering the current state of the industrial society and the importance of Ecolinguistic Learning Modules in providing learners with means to prepare for inevitable changes in the future.

Starting with Rethinking and Reframing Sustainability

Sustainability beyond the 'Three Pillars'

Sustainability-related issues are at the core of ecolinguistic research and education. Therefore, the meaning and framing of sustainability and sustainable development need to be considered when developing Ecolinguistic Learning Modules. Sustainable development is usually described as having three pillars—economic, social, and environmental (e.g., Giddings et al. 2002). The economic pillar standardly refers to economic growth, competitiveness, profit, wealth, and/or prosperity. The three pillars are sometimes expressed more bluntly using an entirely financial framing as the 'triple bottom line' of people, planet, and profit (see e.g., Frisk 2020). Both of these ways of representing sustainability follow the logic of 'while', i.e., making a profit and growing the economy while protecting the planet and looking after people. This is a potentially damaging framing because one of the root causes of ecological destruction is an obsessive focus on growth and profit, and the pillars wrap the pursuit of profit into the pursuit of sustainability itself, so making lots of money becomes part of a sustainability plan, perhaps the main part. As the Intergovernmental Science-Policy Platform on Biodiversity and Ecosystem Services (IPBES 2022) states, 'goals linked primarily to values of individualism and materialism—defining societal progress as wealth, profit, competition, and growth—are not aligned with future sustainability or social justice'. The economy is, of course, a crucial consideration in sustainability but, it should not be represented as an end in itself. Instead, the two goals should be social and environmental, or people and planet, and the aim is to build an economy which allows both society and the environment to flourish.

As long as their concept of sustainability focuses on intrinsic values and includes consideration of the more-than-human world, learners should be free to explore a diverse range of definitions and understandings of sustainability and come up with their own framework. Instead of indoctrinating learners by encouraging them to reproduce the educator's views, active learning is promoted in Ecolinguistic Learning

Modules through instructional activities which involve learners doing tasks, thinking while doing (Bonwell and Eison 1991), and reflecting on what they are doing (e.g., Loughran 2002).

Using Education for Sustainable Development with Caution

Education for Sustainable Development (ESD) was promoted by UNESCO (2005) in the UN Decade of Education for Sustainable Development 2005–2014, and its role as the enabler for the Sustainable Development Goals (SDGs) was established in the Agenda 2030 Action Plan (UNESCO 2018). Since then, ESD has been incorporated into the educational field in numerous UN member countries (e.g., Rudsberg and Öhman 2010) either as part of their institutional policies or via curricula, which often serve as the official guidelines with which teachers plan their lessons. Because of its dominant role in the educational field as well as in research, various environmental and sustainability education are commonly discussed under the umbrella concept of ESD. There is, however, a stark controversy between the approach used in Ecolinguistic Learning Modules and the political and ethical interpretations implied in ESD for sustainability and sustainable development.

These problems were already highlighted by Kopnina (2014) in an article discussing anthropocentric bias during the time when environmental education started to transition towards ESD. Kopnina (2014) concludes that ESD steers focus away from the core environmental problems and promotes social issues and economic development, and in so doing, may be counterproductive to the aim of environmental education to foster environmentally aware and active citizenship that would lead towards solving wicked environmental problems. Therefore, however convenient it might be to use ESD as a platform for Ecolinguistic Learning Modules, it is best to present ESD as one framework alongside others that highlight environmental conservation and non-anthropocentric ways of approaching nature. This enables learners to engage critically with political, institutional, and bureaucratic agendas of sustainability (see e.g., Selby and Kagawa 2010). Many related policies consider different stakeholders, among them industry representatives, and this practice tends to create an unbalanced situation in how different voices are heard. As an example, Indigenous groups represented only 1% of the consulted groups when Australian Academic Standards in Environment and Sustainability were created, whereas employers' and practitioners' representation was 27% (Phelen et al. 2015).

Another cautionary note on using ESD as an educational framework when embedding Ecolinguistic Learning Modules is that even the most recent transformative approaches (e.g., Weinberg et al. 2020; Maijala et al. 2024) may easily be misused for purposes counteracting the initial, intrinsic aims behind sustainability. A key concern is that the skills (e.g., futures thinking, systems thinking, and problem-solving) promoted in the ESD literature (e.g., Brundiers and Wiek 2017; Tarrant and Thiele 2016), while undeniably being useful in working towards sustainability, are so generic that they could equally be used to work towards other aims, such as increasing the profits of ecologically destructive industries. Selby (2007) describes how sustainability skills were framed as generic skills needed by business and industry in a report by Tilbury and others (2004) concerning changes in curricula and graduate skills in the context of Australia.

Despite the controversy, ESD does provide curricular opportunities for Ecolinguistic Learning Modules because it is highlighted by agencies such as the Quality Assurance Agency for Higher Education in the UK (QAA 2021). Using the term sustainability or sustainable development opens doors for modules to be validated which stretch the boundaries of the narrow discipline focus of academia and consider the larger picture of the challenges we face. Hasslöf and Malmberg (2015) suggest that through giving room for *subjectification*, i.e., focusing on personal, responsible actions within a situation and giving learners a possibility to develop a more independent approach, ESD can provide a useful framework for learners to view the world critically.

ESD is discussed in most of the chapters in this volume. In Chapter 2, Niklas Gericke concludes that a multidisciplinary perspective in ESD that relies on contributions from multiple subject-specific teaching traditions would be needed and that the prospects of a transformative language education strategy could be helpful in empowering students' action competence for sustainability. In Chapter 3, Salla-Riikka Kuusalu discusses how the learning processes of ESD can support language learning and how ESD would need to change for it to enhance learning that induces transformative action. Chapter 4 gives a critical view towards the concepts behind ESD, as Attila Krizsán discusses the role of applied linguistics in reconceptualising and changing sustainability-related communication. In Chapter 5, ESD is used as a framework when studying pre-service teachers' future visions of sustainable development's role in language teaching and the pedagogical practices they see as sustainable in two different contexts, and in Chapter 6, Minna Maijala and others underline the importance of implementing ESD in language coursebooks, which function as central tools for language teaching. To conclude, used with caution, ESD can prove helpful for embedding modules in the curriculum, but it is essential that the modules encourage critical reflection (Campbell and Norton 2007) on the assumptions and principles behind the ESD agenda.

Formulating the Theoretical Framework for Embedding Sustainability

Theoretical Basis of Ecolinguistic Learning Modules

The theoretical foundation for the learning modules is derived from critical language awareness (CLA), developed by Norman Fairclough (1992). CLA applies critical discourse analysis (CDA, e.g., Fairclough 1995) to education, helping learners become more active citizens who are aware of the role of discourses in the exploitative social relations around them. As Shapiro (2022) states in their book about how to cultivate CLA in the writing classroom, CLA pedagogy has three main goals: self-reflection, social justice, and rhetorical agency. However, the focus of CLA has been in exposing negative discourses that structure relationships of domination within the human world, leaving out consideration of the larger ecological context. Thus, for developing Ecolinguistic Learning Modules, CLA needed to be reoriented and given a novel context via ecolinguistics.

Ecolinguistics explores the role of language in the life-sustaining interactions of humans, other species and the physical environment. The first aim is to develop linguistic theories which see humans not only as part of society, but also as part of the larger ecosystems that life depends on. The second aim is to show how linguistics can be used to address key ecological issues, from climate change and biodiversity loss to environmental justice. Ecolinguistics as a discipline tends to have a normative orientation towards the protection, preservation, and enhancement of the ecological systems that support life (Fill and Penz 2017), much like medical science, which directs its efforts towards the survival and well-being of our mental and physical bodies. Further, using the ecological analysis of discourses or ecological discourse analysis (EDA, Alexander and Stibbe 2014), one can widen the perspective beyond anthropocentrism and consider how discourses affect other species and the life-supporting ecosystems. Because discourses function at the level of communication among groups of people, ecological discourse analysis usually concentrates on studying discourses of institutions, industries, and social groups that converge around a particular activity.

Another method used in the modules is positive discourse analysis (PDA, Martin 2004; Bartlett 2012, 2017), which helps learners discover inspirational ways of using language that work towards more equal societies and protection of the ecosystems that life depends on. These ways of using language often come from the spoken and oral literature of traditional or Indigenous groups and work towards overcoming oppression by, for example, changing the public opinion which affects decision-making (e.g., Megbowon and Uwah 2020). PDA has also been effective in steering societal values towards a more equitable model, where minority voices are heard and appreciated (Nartey 2020). The foundation for PDA comes from systemic functional linguistics (SFL, see Halliday and Mathiessen 2013), which is a functional theory of the usage of language in social context. As explained by Mallet and Mahboob (2023), SFL concentrates on the main social functions of language, namely representing experience, enacting social relationships and organising these functions, as well as revealing and addressing inequality.

Cognitive linguistics (e.g., Lakoff 2010; Lakoff and Wehling 2012) functions in the learning modules as the foundation for introducing different mental structures expressed in language, such as metaphors and framings, which reveal the stories behind texts. Evaluations, stories representing aspects of life as good or bad, are studied with the help of appraisal theory by Martin and White (2005). Because identity building is a crucial part of a student's development throughout their time in the educational system and beyond, Ecolinguistic Learning Modules put great emphasis on supporting learners to become responsive and responsible citizens who can change their views when realising they no longer serve them or the world that keeps everything alive. Identity theory (Benwell and Stokoe 2006) and its application in the modules is explained in more detail in the section 'Identity in Ecolinguistic Learning Modules'.

Identity in Ecolinguistic Learning Modules

Quilley (2009) describes how education systems tend to assume continuity—the idea that providing learners with the knowledge and skills they need to play a functional role within society will be sufficient to guarantee their future. However, the scale of ecological destruction has worsened significantly (UNEP 2023), which raises profound questions about identity. Because an unsustainable society cannot continue into the future, there are two possibilities: a rapid and total transformation of society into a sustainable one or societal collapse. It is an extremely hard idea to grasp that the world as we know it cannot continue. Sustainability is an example of what Meyer and Land (2005) call a *threshold concept*. Threshold concepts are those which are: *transformative* (occasioning a significant shift in the perception of a subject), *irreversible* (unlikely to be forgotten or unlearned only through considerable effort), and *integrative* (exposing the previously hidden interrelatedness of something). In addition, they may also be *troublesome*. (Meyer and Land 2005, p. 373).

One person who crossed the threshold is Ray Anderson, former CEO of Interface, a large US carpet manufacturing corporation. Anderson was asked in the 1990s to present his environmental vision to staff, but he did not have one. However, after reading Paul Hawken's (1993) book The Ecology of Commerce and realising the meaning of species extinction through Hawken's expression 'the death of birth', he experienced a complete change of heart (Anderson 2005). The result was a deep mental restructuring of his identity: instead of seeing himself as the successful business leader society saw him, he considered himself an environmental criminal. Crossing a threshold like this might make people feel they have invalidated who they are and their life's work. However, this period is only a 'liminal state'—an in-between state between two more solid and comfortable identities (Meyer and Land 2005, p. 374). After his period of self-condemnation, Anderson took on the new identity of a green business leader, made deep changes in his company to make its processes more sustainable, and travelled the world to help other business leaders to transform their companies.

Ecolinguistic Learning Modules can help learners avoid undergoing a painful transition of identity later in life. It is far better for learners to learn identity reflection during their time in education so they can ensure that their emerging professional and personal identities are responsive to and responsible towards the conditions of the world. Instead of uncritically taking on a fixed disciplinary or professional identity, learners may develop more fluid identities, allowing them to respond to changing conditions. They can do this by understanding how society forges identities and the dangers of unquestioningly taking on identities that are offered by an unequal and unsustainable society.

Two kinds of identity theory among others are: essentialist ones which posit a core, authentic identity within each of us (e.g., Fleischer 2005) and postmodern theories which posit multiple, fractured identities that we slip in and out of without any of them being objectively 'us' (Sen 2006). Postmodern theories are most relevant for Ecolinguistic Learning Modules because they allow for the reinvention of self in response to the changing conditions of the world. A fixed identity is necessarily fragile, for it needs to be constantly defended against external attacks or changes which invalidate it, whereas an acceptance of multiple, fluid identities can, paradoxically, contribute to a more stable and grounded sense of self.

Importantly, postmodern perspectives see identities as being 'actively, ongoingly, and dynamically constituted in discourse' (Benwall and Stokoe 2006, p. 4) which is why identity is of vital importance to Ecolinguistic Learning Modules. The modules engage learners in identity reflection, where they critically analyse the ways societal discourses shape identities and make a conscious choice about which identities they want to take on through their own language performance. According to postmodern identity theory (Sen, 2006), education is constantly shaping learners' identities in entirely implicit and unquestioned ways. Learners are often simply informed that certain disciplinary ways are 'correct' and must be performed if they want to graduate as any professional identity. Ecolinguistic Learning Modules encourage learners to become aware of how social forces, including educational institutions, shape their identity and consider ways to resist the ones that feel wrong. This is a form of *political subjectivity*, a means of empowering learners to take a more active role in shaping their identity (Hall 2004). Learners can undertake identity reflection into their future lives and professional career and reforge themselves with the help of deep reflection to better respond to the changing conditions of the world.

The Core Structure of Ecolinguistic Learning Modules

Part 1: Transformational Teaching Steers Learners Towards Critically Engaged Citizenship

Teaching practices of Ecolinguistic Learning Modules lean on the educational framework of transformational teaching (Slavich and Zimbardo 2012), which combines the methods and principles of active, student-centred, collaborative, experiential, and problem-based learning. According to Slavich and Zimbardo (2012), even though these contemporary approaches differ from one another in their literatures, foci, and conceptual scope, they share more similarities than differences, and they are mostly based on similar theoretical roots, such as Vygotskyan (social) constructivism. The core features shared among all five approaches are:

(1) learners learn by doing; (2) teachers are facilitators who provide opportunities for interactions; (3) learner needs are highlighted; and (4) teaching aims at increasing learners' role in the learning experience (Slavich and Zimbardo 2012). Moreover, the objectives of teaching are broader than in conventional teaching, as the aim is not only to teach learners the content of the lecture but as Slavich and Zimbardo (2012, p. 577) aptly state, to 'guide students towards making self-discoveries that shape their fundamental beliefs about themselves'.

Although many ideas that emerge during lessons come from the learners, the teacher frames and guides the entire process, selects the data for the learners, and presents their own ideas, values, and opinions. However, values or worldviews are not forced on the learners, but instead, their critical thinking skills are challenged by offering various viewpoints to consider. Transformational teaching can be challenging for learners who are not familiar with it, but it has more potential to create questioning, critically engaged citizens who interrogate the world around them based on their own deeply reflected-on value system.

Part 2: Learners Explore the Role of Language in Shaping Social Cognition

Ecolinguistic Learning Modules use the concept of ecological philosophy or ecosophy, in this context meaning the framework of values that the learners need to consider before they can proceed to analysing discourses. An ecosophy consists of values related to the interaction of humans with other human beings, as well as with other species, the physical environment, and the ecosystems that support life (Naess 1995; see also Stibbe 2019). To be able to develop their own ecosophy requires that the learners first learn about the multiple ecological philosophies available, such as ecological modernisation (see e.g., Glynn et al. 2017), social ecology (Bookchin 2005), deep adaptation (Bendell and Read 2021), and deep ecology (Naess 1995), to name a few. Learners then select the philosophical system that mostly resonates with them or combine different systems together to form their own ecosophy. Their ecosophy is influenced by direct experience interacting with other people and the natural world and continuously evolves and adapts against time. The ecosophy is crucial for critical evaluation of the stories that surround learners and structure the society. Stories that relate to economic growth, technological progress, the use of natural resources, consumerism, and the natural world are among the most pressing. Once learners have reflected on their values and decided on their own ecosophy, then they use the linguistic tools covered in the classes to analyse texts, revealing the hidden stories and judging them against their own ecosophy. There is a certain amount of knowledge that learners need to learn about linguistic features such as frames, metaphors, presupposition, or transitivity (see Table 11.1). They also need to see worked examples of how analysing these features can reveal the stories we live by.

Part 3: Deep Learning Begins When Learners Put the Theory into Practice

The most important methods in the learning modules are active learning (Bonwell and Eison 1991) and problem-based learning (Graaff and Bouhuijs 1993; Barrett and Moore 2010). For the first active learning exercises, texts that have a clear and representative story to reveal and which are familiar to the educator should be chosen. After the learners are familiar with the exercise, they can choose their own texts around themes they are passionate about. This pedagogical approach combines the aspects of knowledge transmission of linguistic features and analysis techniques with constructionism, which allows learners to discover using their existing knowledge and methods of inquiry. However, because according to Bowers (2005), it is likely that the learners' knowledge is derived from or affected by the dominant cultural discourses that tend to be unjust and unsustainable, constructivism is not enough. Instead, learners need to be exposed to works criticising the status quo, such as the articles about poverty and globalisation (Shiva 2000), The Dark Mountain Project (Kingsnorth and Hine 2009), myths (Midgley 2011), capitalism and climate (Klein 2014), and a living economy (Korten 2015), while also analysing these works with equally careful criticality.

Table 11.1 Some illustrative examples of the kinds of texts and linguistic features that learners explore in Ecolinguistic Learning Modules. Note that all linguistic features could be investigated in all text types

Linguistic Feature	Text Type	What learners are looking for
Activation	Creation stories	Which characters in creation stories are represented as actively contributing to the creation of the world
Lexical sets	Nature writing	How lexical sets associated with magic, mystery, and the supernatural convey a sense of enchantment with ordinary nature
Appraisal patterns	Newspapers	How increases in retail sales are represented as positive and decreases as negative
Presuppositions	Economics textbooks	How humans are represented as unquestionably selfish
Framings	Political speeches	How climate change is framed in alternative ways such as an *environmental problem* or a *security threat*
Metaphor	Environmental writing	How biodiversity is represented metaphorically, e.g., as a *burning library of species* or a *web of life*
Modality	Documentaries	How the description that 'climate change is caused by humans' is represented as true, uncertain, or false
Erasure	Environment reports	How animals, plants, and the natural world are erased through being represented as resources for exploitation
Salience	Animal rights websites	How animals and their lives, deaths, and sufferings are vividly portrayed
Semantic inclusion	Nature poetry	How humans, animals, and plants are represented with a degree of equality through inclusion in the same semantic groupings

In the modules, learners are either working individually, with a pair, in groups, or the whole class together. They read and write, discuss and debate, listen and talk throughout the exercises, depending on the task in question. Bonwell and Eison (1991, p. 5) highlight engagement 'in such higher-order thinking tasks as analysis, synthesis, and evaluation' to actively involve the learners. Therefore, active learning is pronounced in the learning modules to enhance learning (see Barrett and Moore 2010) and because of the nature of ecolinguistic analysis: learners must both do things (i.e., analysis) and think about what they are doing (i.e., reflecting on the ecosophy they are basing the analysis on). Further, learners can experience culture in more ways than their educators may have ever been able to, such as via television, video and console games, social media networks, popular writing, cartoons, streaming services, chat rooms, and AI programmes, to mention a few. These sources are likely to have been overlooked by ecolinguistic researchers, which raises opportunities for even the first-year students to explore their potential as researchers by publishing their findings on e.g., the Ecolinguistic Association website, which functions as a material bank for other researchers globally.

Part 4: Reflective Practices Support the Learners' Identity Development

As stated in 'Introduction', the development of responsive and responsible identities is supported by the reflective practices (e.g., Campbell and Norton 2007) in Ecolinguistic Learning Modules, and identity theories related to the practices are introduced in section 'Identity in Ecolinguistic Learning Modules'. However, reflective practices are part of a wider concept of ecologically responsive moral education, which includes both formal and informal ways in how the moral dispositions of learners are selectively encouraged (Fesmire 2012). By profoundly reviewing and comparing the philosophy traditions of the Eastern Kyoto School and Western (American) pragmatism, Fesmire (2012) showed the importance of relational thinking and ecological imagination in education to develop mental mechanisms that enable actions towards ecological stability and for counteracting and preventing wicked environmental and social problems. Imagination is vital in achieving this because imagining things or events creates the same effect in the brain as experiencing them (see Gibbs 2005), which again is essential because 'It is a general truth that we cannot respond to what we do not perceive, and we will not respond to perceptions unless they are immediately felt'. (Fesmire

2012, p. 121). Further, when we become aware of the relational connections around everything we experience, it amplifies the meanings we give to these experiences, as: '[s]uch meaning enables intelligent and inclusive foresight of the consequences of alternative choices and policies' (Fesmire 2012, p. 120). In other words, developing the competencies of understanding complex (ecological and social) systems and being able to imagine future scenarios and possibilities in relation to current situations and chosen actions, i.e., our *ecological imagination*, is required to support the development of learners' identities.

Hence, on Ecolinguistic Learning Modules, learners engage in identity reflection, which can be structured around a series of questions. The first question is 'Who are you?' and learners will typically respond to this with their name. If pressed further, they are likely to add the fact that they are a student and the subject they are studying: 'An English literature student'. To go further, learners can be encouraged to consider the different identities they adopt in different physical spaces. When out on their bicycle they are a cyclist, when in a hospital they are a patient, when in the mall they are a shopper, and when on a plane they are a passenger. This is the *spatial division of identity* where who we are depends on where we are. To go further, learners can explore *relational identities*—to their brother they are a sister, to their mother they are a daughter, and to their teacher they are a pupil.

The second question is 'How do I perform these identities?', and the answer will involve ways of speaking, writing, dressing, posture, and gestures—people might speak more politely in the identity of pupil than that of sister. The third question is 'Who am I within an unequal and unsustainable society that is on a path towards collapse?'. This is something to think about for all the learners' identities, and there are two basic positions: someone who is working towards social change for a more equal and sustainable society or someone who is contributing to the reproduction of our current ecologically destructive society. The fourth question is 'Who do I feel I need to become?'. For this, the learners may try to say, 'I'm happy as I am', but there is no choice but to change as they continue living their lives. At some point, they will enter a profession and have no choice other than to perform a particular professional identity, as a teacher, lawyer, doctor etc. The aim of this exercise is not to try to force learners into being a particular kind of person but to encourage them to reflect on who they want to become within the real world as it is, i.e., our unsustainable and unequal society. Learners can decide who they want to be, but it is essential that they consider both society and the ecosystems that life depends on in their decision, rather than just passively being forced by society into certain identities.

The fifth question is 'What kind of skills do I need to gain to perform the kind of identities I want to perform in the future?'. For this, learners may want to gain communication skills, analysis skills, writing skills, reflective skills, skills in systems and future thinking, and skills in compassionate decision-making that considers the interests of all stakeholders, including non-human species and future generations. The sixth question is 'Is my current educational institution providing me with the skills I need to perform the identities I want to perform?' The chances are that the answer to this question is 'No' because few educational institutions are fully orientated towards producing graduates who will contribute to a more equal and sustainable society. The learners may then demand educational change or seek educational opportunities outside of their institution to gain the kind of skills they need to be the person they want to be. This is not a one-off exercise but something that learners will need to do repeatedly to avoid having to go through a painful transition when they realise their identity is out of step with the world as it is.

Ecolinguistic Learning Modules are Applicable to Multiple Contexts

Ecolinguistic Learning Modules can be organised around themes that emerge from the surrounding world or around linguistic theory, such that issues of global importance are embedded in the form of case studies. The latter model is more effective in conveying linguistic principles, but it is not entirely satisfactory because it conveys the message that the purpose of education is to learn about the discipline, rather than using the discipline to learn about the world. A truly transdisciplinary approach would start with world issues and bring linguistic and other theories to bear where they are most useful, but it is hard to implement this within the traditional discipline-based education system.

The principles behind the approach are straightforward. They can be incorporated smoothly into dedicated ecolinguistic modules or any linguistics module which explores the relationship between language and the world (e.g., discourse analysis, language and gender, or sociolinguistics). To summarise, the principles for educators of how to work with Ecolinguistic Learning Modules are listed below:

1. Start with guiding the learners to consider the stories society is built on and how injustice and unsustainability call those stories into question.
2. Encourage the learners to reflect on their values and build their own ecosophy for judging the stories against. An ecosophy is a values framework that considers humans, other species, and the ecosystems that life depends on.
3. Teach the learners skills in techniques of linguistic analysis (e.g., critical discourse analysis, cognitive linguistics, identity theory, rhetorical theory) which help them reveal the stories we live by.
4. Guide the learners to use their skills in analysing texts from a wide range of areas of life (e.g., economics textbooks, lifestyle magazines, environmental reports, nature writing) and judging the stories they reveal against their ecosophy.
5. Direct the learners to reflect on how language makes them who they are and towards the language skills they must acquire to become the person they want to be. Encourage them to think about their role as ethical future leaders within an unsustainable society.
6. Help learners to gain skills in oral presentation, new media literacy, persuasive writing, visual communication, and group facilitation so that they can effectively convey new stories to live by as ethical leaders in the future.

The learning modules have been developed originally for the discipline of English Language and Linguistics. However, they have been incorporated with positive results into lectures for staff in business studies, art, classics, media studies, cultural studies, veterinary sciences, environmental sciences, and psychology. The reason for the general applicability of the approach is that all disciplines are based on, deal with, analyse or produce texts. The techniques of linguistic analysis can be used by learners in any discipline to interrogate the texts they work with, reveal the underlying stories, and judge them from a social and ecological perspective. This includes natural sciences since scientific texts encode stories about the nature of reality and about relationships between humans and the more-than-human world in the same way that any other text does.

An example of a ready Ecolinguistic Learning Module is *The Stories We Live By* free online course provided by the University of Gloucestershire and the International Ecolinguistic Association (available at: https://www.storiesweliveby.org.uk/). This course has been taken by more than 4000 participants, and the materials in it have been incorporated into many ecolinguistic modules taught around the world. According to participant feedback, the course has helped in, for example, research framing, understanding ecolinguistic, seeing linguistic features in a new light, using ecosophy as a tool in academic work, and understanding the effectiveness and importance of language in dealing with environmental issues.

Conclusion

This chapter described ecolinguistic as a search for new stories to live by to bring into being a sustainable civilisation. However, there are voices including the artists and writers of the Dark Mountain Project (Kingsnorth and Hine 2009) who argue that it is now too late for that. They argue that the stories of extractivism (for definitions, see Hagolani-Albov 2022), consumerism, and economic growth are far too deeply entrenched across the globe to allow for the kind of total and immediate transformation necessary to avoid civilisational collapse.

There is no reason to assume, according to the Dark Mountain Project Manifesto, that the current pattern of life can or will continue, and plenty of evidence that it cannot for very much longer. In fact, it has always been an inescapable aspect of the concept of sustainability that current societies in developed countries are unsustainable, meaning that they cannot continue indefinitely on the path that they are currently on. The important question is whether the end of industrial civilisation will involve a painless transition to a more sustainable form of civilisation, social and ecological collapse, or, perhaps, a mixture of both, to different degrees in different locations.

Whatever the future holds, it is abundantly clear that education cannot be focused, as so many official policy documents demand, on preparing learners to take their place within an industrial society and contribute to further industrialisation, further economic growth, financial prosperity, international competitiveness, and the production and consumption of even more goods and services. Instead, if the ultimate implication of *unsustainability* is considered, education must now be a preparation for what comes next. Post collapse, priorities may shift to such things as the preservation of vital knowledge from the past, the creation of pockets of resilience in the face of economic, climatic, and ecological disruption, and, for those who remain, the forging of new ways to live based on very different principles.

If learners are not empowered to reflect on their identities throughout their education journey, then all education is doing is preparing them to unquestioningly take up subject positions within an industrialised society on a path towards ecological collapse and thereby contribute towards that collapse, rather than working towards ongoing appraisal and reinvention of self and society. The process of searching for new stories to live by could not be more urgent, either to transform current unsustainable civilisation into a new ecological civilisation or if it is too late for that, then to build a new ecological civilisation after collapse. Ecolinguistic Learning Modules provide essential preparation for both scenarios, helping learners prepare for what comes next and be ready to build a new sustainable society, whatever the future holds.

References

Alexander, R., and A. Stibbe. 2014. From the analysis of ecological discourse to the ecological analysis of discourse. *Language Sciences* 41 (January): 104–110. https://doi.org/10.1016/j.langsci.2013.08.011.

Anderson, J. 2004. The ties that bind? Self- and place-identity in environmental direct action. Ethics. *Place and Environment* 3 (1–2): 45–57. https://doi.org/10.1080/1366879042000264769.

Anderson, R. 2005. *Transcript of interview appearing in the documentary The Corporation.* Vancouver: Big Picture Media.

Augoustinos, M., I. Walker, and N. Donaghue. 2014. *Social cognition: An integrated introduction*, 3rd ed. London: SAGE Publications Ltd.

Barrett, T., and S. Moore. 2010. *New approaches to problem-based learning: Revitalising your practice in higher education.* New York: Routledge.

Bartlett, T. 2012. *Hybrid voices and collaborative change: Contextualising positive discourse analysis.* London: Routledge.

Bartlett, T. 2017. Positive discourse analysis. In *The Routledge handbook of critical discourse studies*, ed. J. Flowerdew, and J.E. Richardson. 1st Ed. Routledge. https://doi.org/10.4324/9781315739342.

Bendell, J., and R. Read. 2021. *Deep adaptation: Navigating the realities of climate chaos*, 1st ed. Cambridge, UK Medford, MA: Polity.

Benwell, B., and E. Stokoe. 2006. *Discourse and identity.* Edinburgh: Edinburgh University Press.

Bookchin, M. 2005. *The ecology of freedom: The emergence and dissolution of hierarchy*. Oakland, CA: AK Press.

Bonwell, C., and J. Eison. 1991. *Active learning: Creating excitement in the classroom.* Washington, DC: School of Education and Human Development, George Washington University.

Bowers, C. 2005. *The false promises of constructivist theories of learning: A global and ecological critique.* New York: Peter Lang.

Brundiers, K., and A. Wiek. 2017. Beyond interpersonal competence: Teaching and learning professional skills in sustainability. *Education Sciences* 7(1) (March 7): 39.

Bürgener, L., and M. Barth. 2018. Sustainability competencies in teacher education: Making teacher education count in everyday school practice. *Journal of Cleaner Production* 174 (February): 821–826. https://doi.org/10.1016/j.jclepro.2017.10.263.

Campbell, A., and L. Norton. 2007. *Learning, teaching and assessing in higher education: Developing reflective practice.* Exeter: Learning Matters.

Fairclough, N. 1992. *Critical language awareness.* London: Longman.

Fairclough, N. 1995. *Critical discourse analysis.* London: Longman.

Fesmire, S. 2012. Ecological imagination and aims of moral education through the Kyoto school and American pragmatism. In *Education and the Kyoto School of Philosophy*, ed. P. Standish, and N. Saito, 109–130. Dordrecht: Springer Netherlands. https://doi.org/10.1007/978-94-007-4047-1_9.

Fill, A., and H. Penz. 2017. *The Routledge handbook of ecolinguistic*, 1st ed. New York, NY: Routledge.

Fleischer, L. 2005. The development of authentic identity: Implications for the soul of education. *Reclaiming Children and Youth* 14 (3): 179–183.

Frisk, P. 2020. *People planet profit: How to embrace sustainability for innovation and business growth.* London: Kogan Page Publishers.

Frisk, E., and K.L. Larson. 2011. Educating for sustainability: Competencies & practices for transformative action. *Journal of Sustainability Education* 2, no. 1 (March): 1–20.

Gibbs, R. 2005. *Embodiment in cognitive science.* New York: Cambridge University Press.

Giddings, B., B. Hopwood, and G. O'Brien. 2002. Environment, economy and society: Fitting them together into sustainable development. *Sustainable Development* 10, no. 4 (November): 187–196. https://doi.org/10.1002/sd.199.

Glynn, P., T. Cadman, and T. Maraseni. 2017. Ecological modernization: Theory and the policy process. In *Business, organized labour and climate policy*, 22–46. Edward Elgar Publishing. Available from: https://www.elgaronline.com/view/9781786430113/chapter01.xhtml. Accessed October 2, 2024.

Graaff, E., and P. de Bouhuijs. 1993. *Implementation of problem-based learning in higher education.* Amsterdam: Thesis Publishers.

Hagolani-Albov, S. 2022. Extractivism. In *Showing theory to know theory: Understanding social science concepts through illustrative vignettes*, ed. P. Ballamingie, and D. Szanto. Showing Theory Press. Available from: https://ecampusontario.pressbooks.pub/showingtheory/chapter/extractivism/. Accessed October 2, 2024.

Hall, D. 2004. *Subjectivity.* London: Routledge.

Halliday, M.A.K., and C.M.I.M. Matthiessen. 2013. Halliday's introduction to functional grammar. *Routledge*. https://doi.org/10.4324/9780203431269.

Hasslöf, H., and C. Malmberg. 2015. Critical thinking as room for subjectification in education for sustainable development. *Environmental Education Research* 21, no. 2 (February 17): 239–255. https://doi.org/10.1080/13504622.2014.940854.

Hawken, P. 1993. *The ecology of commerce: A declaration of sustainability*. New York: HarperCollins Publisher.

IPBES. 2022. Intergovernmental science-policy platform on biodiversity and ecosystem services. *Summary for policymakers of the methodological assessment of the diverse values and valuation of nature of the Intergovernmental Science-Policy Platform on Biodiversity and Ecosystem Services (IPBES)* (version 1.2). Zenodo. Available from: https://zenodo.org/record/6522392. Accessed November 13, 2024.

Kingsnorth, P. and D. Hine. 2009. *The dark mountain project manifesto* [online]. Available from: http://dark-mountain.net/about/manifesto/. Accessed October 2, 2024.

Klein, N. 2014. *This changes everything: Capitalism vs. the climate*. New York: Simon and Schuster.

Korten, D. 2015. *Change the story, change the future: A living economy for a living earth.* Oakland, CA: Berrett-Koehler.

Kopnina, H. 2014. Revisiting Education for Sustainable Development (ESD): Examining anthropocentric bias through the transition of environmental education to ESD. *Sustainable Development* 22, no. 2 (March): 73–83. https://doi.org/10.1002/sd.529.

Lakoff, G. 2010. Why it matters how we frame the environment. *Environmental Communication: A Journal of Nature and Culture* 4 (1): 70–81.

Lakoff, G., and E. Wehling. 2012. *The little blue book: The essential guide to thinking and talking democratic*. New York: Free Press.

Loughran, J. John. 2002. Effective reflective practice: In search of meaning in learning about teaching. *Journal of Teacher Education* 53 (1): 33–43. https://doi.org/10.1177/0022487102053001004.

Lundegård, I., and P.O. Wickman. 2007. Conflicts of interest: An indispensable element of Education for sustainable development. *Environmental Education Research* 13 (1): 1–15. https://doi.org/10.1080/13504620601122566.

Lundegård, I., and P.O. Wickman. 2012. It takes two to tango: Studying how students constitute political subjects in discourses on sustainable development. *Environmental Education Research* 18 (2): 153–169. https://doi.org/10.1080/13504622.2011.590895.

Maijala, M., N. Gericke, S.-R. Kuusalu, L.M. Heikkola, M. Mutta, K. Mäntylä, and J. Rose. 2024. Conceptualising transformative language teaching for sustainability and why it is needed. *Environmental Education Research* 1–20. https://doi.org/10.1080/13504622.2023.2167941.

Mallet, A., and A. Mahboob. 2023. Positive Discourse Analysis. In *Linguistics*, ed. A. Mallet, and A. Mahboob. Oxford University Press. Available from: https://oxfordbibliographies.com/display/document/obo-9780199772810/obo-9780199772810-0306.xml. Accessed September 18, 2024.

Manuele, F.A. 2015. Culture change agent: The overarching role of OSH professionals. *Professional Safety* 60, no. 12: 38–44. Available from: https://www.jstor.org/stable/48691105. accessed November 13, 2024.

Martin, J. 2004. Positive discourse analysis: Solidarity and change. *Revista Canaria De Estudios Ingleses* 49: 179–200.

Martin, J., and P. White. 2005. *The language of evaluation: Appraisal in English*. New York: Palgrave Macmillan.

Megbowon, F.K., and C. Uwah. 2020. African literature: A tool for societal regeneration. *Imbizo* 11, no. 1 (May 12). Available from: https://www.upjournals.co.za/index.php/Imbizo/article/view/6432. Accessed September 18, 2024.

Meyer, J., and R. Land. 2005. Threshold concepts and troublesome knowledge (2): Epistemological considerations and a conceptual framework for teaching and learning. *Higher Education* 49 (3): 373–388. https://doi.org/10.1007/s10734-004-6779-5.

Midgley, M. 2011. *The myths we live by*. New York: Routledge.

Miller, E.R., H. Kayi-Aydar, M. Varghese, and G. Vitanova. 2018. Editors' introduction to interdisciplinarity in language teacher agency: Theoretical and analytical explorations. *System* 79 (December): 1–6. https://doi.org/10.1016/j.system.2018.07.008.

Naess, A. 1995. The shallow and the long range, deep ecology movement. In *The deep ecology movement: An introductory anthology*. ed. A. Drengson, and Y. Inoue. Berkeley: North Atlantic Books, 3–10.

Nartey, M. 2020. Voice, agency and identity: A positive discourse analysis of "resistance" in the rhetoric of Kwame Nkrumah. *Language and Intercultural Communication* 20, no. 2 (March 3): 193–205. https://doi.org/10.1080/14708477.2020.1722686.

Odell, V., P. Molthan-Hill, S. Martin, and S. Sterling. 2019. Transformative education to address all sustainable development goals. In *Quality education. Encyclopedia of the UN Sustainable Development Goals*, 905–916. Cham, Switzerland: Springer.

Öhman, J. 2006. Pluralism and criticism in environmental education and education for sustainable development: A practical understanding. *Environmental Education Research* 12 (2): 149–163. https://doi.org/10.1080/13504620600688856.

Osberg, D., and G. Biesta. 2010. The end/s of education: Complexity and the conundrum of the inclusive educational curriculum. *International Journal of Inclusive Education* 14 (6): 593–607. https://doi.org/10.1080/13603110802530684.

Phelen, L., B. McBain, A. Ferguson, P. Brown, V. Brown, I. Hay, R. Horsfield, and R. Taplin. 2015. *Learning and teaching academic standards statement for environment and sustainability* [online]. Sydney: Office for Learning and Teaching. Available from: http://www.researchgate.net/publication/274006942_Learning_and_teaching_academic_standards_statement_for_environment_and_sustainability. Accessed September 18, 2024.

QAA. 2021. *Education for sustainable development guidance. Executive summary.* London: The Quality Assurance Agency for Higher Education and Advance HE.

Quilley, S. 2009. Transition skills. In *The handbook of sustainability literacy*, ed. A. Stibbe, and A. Dartington: Green Books, 43–51.

Rudsberg, K., and J. Öhman. 2010. Pluralism in practice—Experiences from Swedish evaluation, school development and research. *Environmental Education Research* 16, no. 1 (February): 95–111.

Scott, W., and S. Gough. 2003. *Sustainable development and learning: Framing the issues*. London: Routledge Falmer.

Shapiro, S. 2022. *Cultivating critical language awareness in the writing classroom*, 1st ed. New York: Routledge. https://doi.org/10.4324/9781003171751.

Shiva, V. 2000. *Poverty and globalisation* [online]. BBC online network. Available from: http://news.bbc.co.uk/hi/english/static/events/reith_2000/lecture5.stm. Accessed October 2, 2024.

Selby, D. 2007. As the heating happens: Education for sustainable development or education for sustainable contraction? *International Journal of Innovation and Sustainable Development.* 2 (3): 249–267. https://doi.org/10.1504/IJISD.2007.017938.

Selby, D., and F. Kagawa. 2010. Runaway climate change as challenge to the 'closing circle' of education for sustainable development. *Journal of Education for Sustainable Development.* 4 (1): 37–50. https://doi.org/10.1177/097340820900400111.

Sen, A. 2006. *Identity and violence: The illusion of destiny*. London: Allen Lane.

Slavich, G.M., and P.G. Zimbardo. 2012. Transformational teaching: Theoretical underpinnings, basic principles, and core methods. *Educational Psychology Review 24*, no. 4 (December): 569–608.

Steffensen, S.V., and A. Fill. 2014. Ecolinguistic: The state of the art and future horizons. *Language Sciences* 41 (January): 6–25. https://doi.org/10.1016/j.langsci.2013.08.003.

Stibbe, A. 2015. *Ecolinguistic: Language, ecology and the stories we live by*. London: Routledge.

Stibbe, A. 2019. Education for sustainability and the stories we live by. In *Prioritizing sustainability education: A comprehensive approach*, ed. J. Armon, S. Scoffham, and C. Armon. 1st Published. Abingdon, Oxon New York, NY: Routledge.

Tarrant, S.P., and L.P. Thiele. 2016. Practice makes pedagogy—John Dewey and skills-based sustainability education. *International Journal of Sustainability in Higher Education* 17, no. 1 (January 4): 54–67.

Tilbury, D., D. Podger, and A. Reid 2004. *Change in curricula and graduate skills towards sustainability: Final report*. Sydney: Australian Government Department of the Environment and Heritage/Macquarie University.

UNEP. 2023. *Global environmental outlook. United nations environment program*. Available from: https://www.unep.org/annualreport/. Accessed September 20, 2024.

UNESCO. 2005. *UN decade of education for sustainable development 2005–2014. The DESD at a glance*. United Nations Educational, Scientific and Cultural Organization [online]. Available from: https://unesdoc.unesco.org/ark:/48223/pf0000141629. Accessed September 20, 2024.

UNESCO. 2018. *Global action programme on education for sustainable development. Education for sustainable development and the SDGs. Policy brief: Advancing ESD policy.* United nations educational, scientific and cultural organization [online]. Available from: https://unesdoc.unesco.org/ark:/48223/pf0000246270. Accessed July 22, 2024.

Weinberg, A.E., C.D. Trott, W. Wakefield, E.G. Merritt, and L. Archambault. 2020. Looking inward, outward, and forward: Exploring the process of transformative learning in teacher education for a sustainable future. *Sustainability Science* 15, no. 6 (November): 1767–1787. https://doi.org/10.1007/s11625-020-00831-9.

Index

M. Maijala et al. (eds.), *Transformative Language Teaching for Sustainability*,
https://doi.org/10.1007/978-3-031-85493-4

The manufacturer's authorised representative in the EU is Springer Nature Customer Service Centre GmbH, Europaplatz 3, 69115 Heidelberg, Germany. If you have any concerns regarding our products, please contact ProductSafety@springernature.com

Printed and bound by CPI Group (UK) Ltd, Croydon, CR0 4YY
01/10/2026
02225953-0012